**Integral Equations and
Stability of Feedback Systems**

This is Volume 104 in
MATHEMATICS IN SCIENCE AND ENGINEERING
A series of monographs and textbooks
Edited by RICHARD BELLMAN, *University of Southern California*

The complete listing of books in this series is available from the Publisher
upon request.

Integral Equations and Stability of Feedback Systems

CONSTANTIN CORDUNEANU

University of Jassy
Jassy, Romania

ACADEMIC PRESS   New York and London   1973

A Subsidiary of Harcourt Brace Jovanovich, Publishers

ACADEMIC PRESS, INC.
111 Fifth Avenue, New York, New York 10003

United Kingdom Edition published by
ACADEMIC PRESS, INC. (LONDON) LTD.
24/28 Oval Road, London NW1

Library of Congress Cataloging in Publication Data

Corduneanu, C
    Integral equations and stability of feedback systems.

    Bibliography: p.
    1.   Control theory.   2.   Feedback control systems.
3.   Integral equations.   I.   Title.
QA402.3.C66   629.8'312   72–88341
ISBN 0–12–188350–7

AMS (MOS) 1970 Subject Classifications: 45M10, 45D05,
45M05, 45E10

Contents

Preface

The mathematical description of the dynamical processes encountered in physical, biological, and applied sciences nowadays requires the investigation of (ordinary and partial) differential equations, equations with delay, integral or integro–differential equations, and other kinds of functional equations.

The first integral equation mentioned in the mathematical literature is due to Abel and can be found in almost any book on this subject (see, for instance, the author's book [12]). Abel found this equation in 1812, starting from a problem in mechanics. He gave a very elegant solution that was published in 1826 in *Crelle's Journal*.

Starting in 1896, Vito Volterra built up a theory of integral equations, viewing their solution as a problem of finding the inverses of certain integral operators.

In 1900, Ivar Fredholm made his famous contribution that led to a fascinating period in the development of mathematical analysis. Poincaré, Fréchet, Hilbert, Schmidt, Hardy, and Riesz were involved in this new area of research. The impact of Fredholm's theory on the foundation and development of functional analysis has also been outstanding. These facts could explain why Volterra's equations, whose role in the investigation of some dynamical processes (mainly in biology) had been emphasized by Volterra himself, took a place of secondary importance. Actually, the integral equations of Volterra type are present any time we deal with a differential equation. According to J. Dieudonné, "... differential equations constitute a swindling. In fact, there exist no differential equations. The only interesting equations are the integral ones" (*Nico*, 1969, No. 2).

During the last 15–20 years, Volterra equations have emerged vigorously in such applied fields as automatic control theory, network theory, and the dynamics of nuclear reactors. The fact that causal operators (i.e., Volterra operators) alone are adequate to describe dynamical systems has led to an increased interest in Volterra equations. Satisfactory solutions have been found for nonlinear problems, which very often arise in applications. Many results are now available, and several books on this subject have appeared in the last few years (see Kudrewicz [5], Miller [7], and J. C. Willems [3]).

Except for the introductory chapter and a few results in Chapters 2 and 4, this book contains concepts and results related to Volterra equations. Some topics are not necessarily related only to Volterra equations and, therefore, a broader framework has been considered in the exposition.

Chapter 2 is concerned with linear integral operators (admissibility results) and Hammerstein equations. The first part of the chapter contains necessary and sufficient conditions for the continuity of various classes of linear integral operators. The reader interested in applications will recognize here various criteria of stability for linear input–output systems. The second part of the chapter is devoted to the existence theory for Hammerstein equations, using fixed-point theorems as the main tool.

Chapter 3 contains results pertaining to the frequency method. Besides Popov's original stability theorem, there are some generalizations and related topics. Applications are also included.

Chapter 4 is dedicated to the theory of the Wiener–Hopf equation, as it appears in M. G. Krein's work [1]. The underlying space here is richer than in Krein's original theory. This theory has significant applications (the transfer of radiant energy, for instance) that have not been included in this book.

The last chapter is heterogeneous in character and contains various results related to the use of Liapunov functionals (energy method), positive kernels, the linearization problem, and Volterra–Stieltjes equations, as well as an application to the dynamics of nuclear reactors (due to Levin and Nohel [6]).

The applied mathematician and the research engineer will find in this book some tools of current use in systems theory. The pure mathematician can see how concepts and results such as Banach algebras, the closed-graph theorem, or fixed-point theorems are used to reach very profane purposes.

Many results included in this book have been discussed at the weekly meetings of the seminar on qualitative problems with the Faculty of Mathematics, University of Iaşi, during the academic year 1966/1967 and since

1968. A seminar on integral equations was held in the Department of Mathematics at the University of Rhode Island during the academic year 1967/1968, while the author held a visiting position there.

The author is indebted to Professor D. Petrovanu, Professor R. K. Miller, Dr. V. Barbu, and Dr. G. Bantaş for helpful discussions. In particular, Dr. Barbu cooperated in writing Section 4.4. We take this opportunity to express our sincere thanks.

We also wish to thank Miss Roza Oxenberg for her cooperation in preparing the manuscript.

Finally, our thanks and appreciation go to Professor R. Bellman, and to all persons at Academic Press involved in carrying out this project.

Preliminaries

This chapter provides some auxiliary results we shall need in subsequent chapters. Because many of these results can be easily found in the mathematical literature, we do not include their proofs. Only a sketch of the proof and adequate references are given. There are some results for which we give the complete proofs. We intend to spare the reader's effort in better understanding some theorems that play an important role in our exposition and whose interest overreaches the framework of this book.

1.1 Some Function Spaces

We shall deal first with some spaces of continuous functions on the positive half-axis $R_+ = \{t : t \geq 0\}$. Following Bourbaki [1], let us denote by $C_c(R_+, R^n)$ the linear space of all continuous mappings from R_+ into R^n, the topology being that of uniform convergence on any compact subset of R_+.

For any $x \in C_c(R_+ , R^n)$ and natural number $n \geq 1$, denote

$$|x|_n = \sup\{\|x(t)\| : 0 \leq t \leq n\}, \qquad (1.1)$$

where $\|\cdot\|$ stands for the Euclidean norm in R^n. The mapping $x \to |x|_n$ is a seminorm on $C_c(R_+ , R^n)$ and the topology of uniform convergence on any compact subset of R_+ is that generated by the family of seminorms $\{|x|_n : n = 1, 2, \ldots\}$. A distance function can be defined on $C_c(R_+ , R^n)$ by means of the formula

$$\rho(x, y) = \sum_{n=1}^{\infty} \frac{1}{2^n} \frac{|x - y|_n}{1 + |x - y|_n}. \qquad (1.2)$$

The topology induced by the distance function ρ is the same topology of uniform convergence on any compact subset of R_+. It can be easily shown that $C_c(R_+ , R^n)$ is complete.

Finally, let us mention that a family of functions $M \subset C_c(R_+ , R^n)$ is relatively compact if and only if the following property holds: let $T > 0$ be an arbitrary number and consider the set of restrictions of the functions belonging to M on $[0, T]$, say M_T; then M_T is relatively compact in $C([0, T], R^n)$, i.e., it is uniformly bounded and equicontinuous on $[0, T]$.

Consider now a continuous positive function $g(t)$, $t \in R_+$. By $C_g(R_+ , R^n)$ we will denote the linear space of all continuous mappings from R_+ into R^n such that

$$\sup\{\|x(t)\|/g(t) : t \in R_+\} < +\infty. \qquad (1.3)$$

If we define

$$|x|_{C_g} = \sup\{\|x(t)\|/g(t) : t \in R_+\} \qquad (1.4)$$

for all $x \in C_g(R_+ , R^n)$, then we can see easily that $x \to |x|_{C_g}$ is a norm.

Moreover, $C_g(R_+ , R^n)$ is a Banach space. To prove this statement we shall consider a Cauchy sequence $\{x^m\} \subset C_g(R_+ , R^n)$, and show that there exists $x \in C_g(R_+ , R^n)$ with the property

$$\lim_{m \to \infty} |x^m - x|_{C_g} = 0. \qquad (1.5)$$

Indeed, to any $\varepsilon > 0$ there corresponds $N = N(\varepsilon) > 0$ such that

$$|x^m - x^p|_{C_g} < \varepsilon \qquad (1.6)$$

whenever $m, p \geq N(\varepsilon)$. But (1.6) implies

$$\|x^m(t) - x^p(t)\| < \varepsilon g(t) \qquad (1.7)$$

for $t \in R_+$ and $m, p \geq N(\varepsilon)$. From (1.7) we easily derive the existence of a mapping $x(t)$ such that $x(t) = \lim x^m(t)$ as $m \to \infty$, for any fixed $t \in R_+$.

Morever, it follows from (1.7) that $x(t) = \lim x^m(t)$ as $m \to \infty$, uniformly on any finite interval of R_+. Hence, $x(t)$ is a continuous mapping from R_+ into R^n. We have to prove only that $x \in C_g$. Let us keep m fixed and make $p \to \infty$ in (1.7). We obtain $x^m - x \in C_g$ for $m \geq N(\varepsilon)$. But $x = (x - x^m) + x^m$ and, since both terms in the right-hand side belong to C_g, it follows that $x \in C_g$.

It is useful to point out that the topology of C_g is stronger than the topology of C_c. Indeed, assume $x^m \to x$ in C_g as $m \to \infty$. Then, given $\varepsilon > 0$, there corresponds $N(\varepsilon) > 0$ such that

$$\|x^m(t) - x(t)\| < \varepsilon g(t), \qquad t \in R_+, \qquad (1.8)$$

whenever $m \geq N(\varepsilon)$. Since $g(t)$ is bounded on any compact interval $[0, T]$, (1.8) shows that $x^m(t) \to x(t)$ as $m \to \infty$, uniformly on any compact subset of R_+. In other words, convergence in C_g implies convergence in C_c.

For $g(t) \equiv 1$ on R_+, the space C_g becomes the well-known space $C = C(R_+, R^n)$ consisting of all continuous bounded mappings from R_+ into R^n, with the topology of uniform convergence on R_+. The norm is then given by

$$|x|_C = \sup\{\|x(t)\| : t \in R_+\}. \qquad (1.9)$$

It is obvious that any C_g is isomorphic with C. The isomorphic mapping from C into C_g is $x \to xg$. Nevertheless, the possibilities in choosing $g(t)$ provide a large variety of spaces, each space consisting of continuous functions with a certain behavior on the half-axis R_+.

Several subspaces of the space C are needed in the sequel. For instance, the mappings from C such that there exists $\lim x(t) = x(\infty)$ as $t \to +\infty$ form a (closed) subspace of the Banach space C. We shall denote it by $C_\ell = C_\ell(R_+, R^n)$. A compactness criterion in C_ℓ will be used later, and the reader can easily formulate it if one takes into account that $C_\ell(R_+, R^n)$ is isomorphic to $C([0, T], R^n)$ where $T > 0$. Indeed, if $x(\tau) \in C([0, T], R^n)$, then $y(t) = x(\tau)$ for $t = \tau(T - \tau)^{-1}$, $\tau \in [0, T)$, defines a function from $C_\ell(R_+, R^n)$.

Another Banach space of continuous mappings from R_+ into R^n is the subspace of C, say C_0, such that $\lim x(t) = 0$ as $t \to +\infty$ for any $x \in C_0$. Of course, C_0 is also a subspace of the space C_ℓ.

For any $\omega > 0$, we shall denote by A_ω the subspace of $C(R_+, R^n)$ consisting of all ω-periodic mappings: $x(t + \omega) = x(t)$, $t \in R_+$.

If we consider the space $C(R, R^n)$, i.e., the Banach space of all continuous bounded mappings from R into R^n, with the topology given by the norm

$$|x|_C = \sup\{\|x(t)\| : t \in R\}, \qquad (1.10)$$

then A_ω can be treated also as a subspace of $C(R, R^n)$.

Denote by $AP = AP(R, R^n)$ the space of almost periodic mappings from R into R^n. Almost periodicity is meant here in the sense of Bohr (see Dunford and Schwartz [1] or C. Corduneanu [8]). Then $AP(R, R^n)$ is a subspace of $C(R, R^n)$. Of course, $A_\omega \subset AP$ for any $\omega > 0$.

Let us consider now some function spaces consisting of measurable functions. The term "measurable" always refers to Lebesgue measure.

By $L_{loc} = L_{loc}(R_+, R^n)$ we denote the space of all measurable mappings from R_+ into R^n such that $x \in L_{loc}$ if and only if $x(t)$ is locally integrable on R_+ (i.e., $x(t)$ is integrable on any compact subset of R_+). As usual, we agree to identify two functions that differ only on a set of measure zero. The topology is generated by the family of seminorms $x \to |x|_n$, $n = 1, 2, \ldots$, where

$$|x|_n = \int_0^n \|x(t)\| \, dt, \qquad n = 1, 2, \ldots. \tag{1.11}$$

Consequently, $L_{loc}(R_+, R^n)$ is a Fréchet space. The reader can easily define the space $L_{loc}(R, R^n)$.

Among the various subspaces of $L_{loc}(R_+, R^n)$, we shall mention first $M = M(R_+, R^n)$. This space consists of all $x(t) \in L_{loc}(R_+, R^n)$ such that

$$\sup\left\{ \int_t^{t+1} \|x(s)\| \, ds : t \in R_+ \right\} < +\infty. \tag{1.12}$$

The topology of M is given by the norm

$$|x|_M = \sup\left\{ \int_t^{t+1} \|x(s)\| \, ds : t \in R_+ \right\}. \tag{1.13}$$

It can be easily proven that M is a Banach space and that its topology is stronger than the topology of $L_{loc}(R_+, R^n)$. Further properties of M can be found in Massera and Schäffer [1]. The meaning of the symbol $M(R, R^n)$ is obvious.

We assume the reader is aware of the theory of Lebesgue spaces $L^p = L^p(R, R^n)$, $1 \le p \le \infty$. Each L^p is a Banach space, with the norm

$$|x|_{L^p} = \left\{ \int_R \|x(s)\|^p \, ds \right\}^{1/p}, \qquad 1 \le p < \infty, \tag{1.14}$$

and

$$|x|_{L^\infty} = \text{ess sup}\{\|x(t)\| : t \in R\}. \tag{1.15}$$

The following relationship between $M(R, R^n)$ and $L^p(R, R^n)$ can be established:

$$L^p(R, R^n) \subset M(R, R^n), \qquad 1 \le p \le \infty. \tag{1.16}$$

For $p = 1$ and $p = \infty$, the inclusion is obvious. If $1 < p < \infty$, then

$$\int_t^{t+1} \|x(s)\| \, ds \le \left\{ \int_t^{t+1} \|x(s)\|^p \, ds \right\}^{1/p}. \tag{1.17}$$

From (1.17) we easily derive that $x \in L^p$ implies $x \in M$. Moreover, we have

$$|x|_M \le |x|_{L_p}, \qquad 1 \le p \le \infty. \tag{1.18}$$

It follows from (1.18) that the topology of L^p, $1 \le p \le \infty$, is stronger than the topology of M.

Since $C(R_+, R^n) \subset L^\infty(R_+, R^n)$ and for any $x \in C$ one has

$$|x|_C = |x|_{L^\infty}, \tag{1.19}$$

it follows immediately that C is a (closed) subspace of L^∞. Therefore, any subspace of C (for instance, C_ℓ, A_ω, C_0) is also a subspace of L.

Among the subspaces of $M(R, R^n)$ we shall mention the space of almost periodic functions in the Stepanov sense (see C. Corduneanu [8]). It will be denoted by $S = S(R, R^n)$ and consists of all mappings $x \in M$ such that the following property holds: to any $\varepsilon > 0$, there corresponds $\ell = \ell(\varepsilon) > 0$ such that any interval $(\alpha, \alpha + \ell)$ of the real axis contains at least one τ for which

$$|x(t + \tau) - x(t)|_M < \varepsilon. \tag{1.20}$$

More specifically, S represents the largest space of almost periodic functions in the Stepanov sense. Indeed, any such function is related to a certain $p \ge 1$, and an almost periodic function of order $p > 1$ is always almost periodic of order 1, i.e., in the sense defined above (see C. Corduneanu [8]).

By P_ω we shall denote the space of locally integrable functions with period $\omega > 0$. In other words, $x \in P$ if and only if $x(t + \omega) = x(t)$ for almost all $t \in R$. The norm in P_ω is that of the space M. It can be easily seen that

$$x \to \int_0^\omega \|x(t)\| \, dt \tag{1.21}$$

is another norm on P_ω, equivalent to the norm of the space M.

The following compactness criterion holds in the space P_ω (see, for instance, Dunford and Schwartz [1, Chapter IV, §8]):

A subset A of P_ω is relatively compact if and only if there exists a $K > 0$ such that

$$\int_0^\omega \|x(t)\| \, dt \le K, \qquad x \in A, \tag{1.22}$$

and

$$\lim_{h \to 0} \int_0^\omega \|x(t + h) - x(t)\| \, dt = 0, \tag{1.23}$$

uniformly with respect to $x \in A$.

Exercises

1. Let us consider the spaces $C_\ell(R_+, R^n)$ and $C_0(R_+, R^n)$. Show that C_ℓ is the direct sum of C_0 and R^n.

2. Consider the space $M(R_+, R^n)$ and show that $x(t) \in L_{\text{loc}}(R_+, R^n)$ belongs to M if and only if for any $\alpha > 0$

$$\sup\left\{\int_0^t \|x(s)\|\, e^{-\alpha(t-s)}\, ds : t \in R_+\right\} < +\infty.$$

Show also that $x(t) \in L_{\text{loc}}(R_+, R^n)$ belongs to M if and only if for any $\alpha > 0$

$$\sup\left\{\int_t^\infty \|x(s)\|\, e^{\alpha(t-s)}\, ds : t \in R_+\right\} < +\infty.$$

Both quantities involved above, regarded as functions of $x \in M$, are norms on M, equivalent to the norm $|x|_M$ (see Massera and Schäffer [1]).

3. Denote by $M_0(R_+, R^n)$ the subspace of $M(R_+, R^n)$ consisting of all $x(t)$ such that

$$\int_t^{t+1} \|x(s)\|\, ds \to 0 \qquad \text{as} \quad t \to +\infty.$$

Show that $L^p(R_+, R^n) \subset M_0(R_+, R^n)$ for $1 \le p < \infty$.

4. Let $M^p(R_+, R^n)$, $1 \le p < \infty$, be the set of all functions in $L_{\text{loc}}(R_+, R^n)$ such that

$$\sup\left\{\left(\int_t^{t+1} \|x(s)\|^p\, ds\right)^{1/p} : t \in R_+\right\} < +\infty.$$

Then M^p is a Banach space with the norm

$$|x|_{M^p} = \sup\left\{\left(\int_t^{t+1} \|x(s)\|^p\, ds\right)^{1/p} : t \in R_+\right\}.$$

Show that $M^p \subset M^1 = M$ for any p, $1 < p < \infty$. Moreover, the topology of M^p, $1 < p < \infty$, is stronger than the topology of M. Discuss similar properties to those encountered in Exercise 2.

5. Let x be an arbitrary function in $M(R_+, R^n)$ and $\alpha > 0$. Show that there exist positive numbers $A(\alpha)$ and $B(\alpha)$ such that

$$A(\alpha)|x|_M \le \sup\left\{\int_t^{t+\alpha} \|x(s)\|\, ds : t \in R_+\right\} \le B(\alpha)|x|_M.$$

6. Consider the spaces $L([0,1], R^n)$ and $M(R_+, R^n)$. Show that M can be identified with the space of all bounded sequences from L, the norm being defined as follows: if $\xi = \{x^m\} \subset L$, then

$$|\xi| = \sup\{|x^m|_L : m = 1, 2, \ldots\}.$$

1.2   A Fixed-Point Theorem

One of the most useful tools in the proof of existence theorems for functional equations is the result usually known as the Schauder–Tychonoff fixed-point theorem.

If S is an arbitrary set and f is a mapping from S into itself, then $x \in S$ is called a fixed point for f if $x = f(x)$. There are many interesting results concerning the existence of fixed points and the reader desiring more information on this subject will find supplementary results in Cronin's book [1].

We shall restrict our considerations to the case when S is a locally convex Hausdorff space (see Yosida [1, Chapter 1, §1]) and f is a continuous mapping satisfying suitable conditions. The proof we shall present below is due to Hukuhara [1].

Several auxiliary results are needed in the proof of the Schauder–Tychonoff fixed-point theorem.

Lemma 2.1   (Brouwer)   Let $\Sigma = \{x : x \in R^n, \|x\| \le \rho, \rho > 0\}$ be a ball in R^n and assume that f is a continuous mapping of Σ into itself. Then there exists at least one element $x \in \Sigma$ such that $x = f(x)$.

The proof can be found in Dunford and Schwartz [1, Chapter V, §12] and in many books on topology.

Lemma 2.2   Let S be a topological space that is homeomorphic to the ball $\Sigma = \{x : x \in R^n, \|x\| \le \rho, \rho > 0\}$. Then, any continuous mapping of S into itself has at least one fixed point.

Proof   Let f be a continuous mapping of S into itself. Assume $\varphi : S \to \Sigma$ is a homeomorphic mapping. If $\xi = \varphi(x)$ and $\eta = \varphi(f(x))$, it follows that $\eta = \varphi(f(\varphi^{-1}(\xi)))$ is a continuous mapping of Σ into itself. If $\bar{\xi}$ is a fixed point of this mapping, then $\bar{x} = \varphi^{-1}(\bar{\xi})$ is a fixed point for f.

Lemma 2.3   Let E be a linear topological Hausdorff space (Yosida, [1, Chapter 1, §1]) of finite dimension. Assume $K \subset E$ is a compact convex set. Then K is homeomorphic to the ball $\Sigma = \{x : x \in R^n, \|x\| \le \rho, \rho > 0\}$, for suitable n.

Proof   Without loss of generality, we can assume $E = R^m$ (see Valentine [1]). If K is a convex body in R^m, i.e., if K has interior points, then we can take $n = m$. If the dimension of K is less than m, then there exists a linear manifold $E_1 \subset R^m$ of minimal dimension n, containing K. Since E_1 is linearly equivalent to R^n, the problem reduces to the case where K is a convex body.

There remains to show that a compact convex body in R^n is homeomorphic to the ball Σ. One can obviously assume that K contains the origin in R^n. The homeomorphic mapping from K onto Σ can be easily constructed. Indeed, let us consider a variable ray starting from the origin of R^n. It cuts the boundary of K at a single point A, and that of Σ at another point, say $\tilde{A}$. To each $x \in OA$ we will associate a unique point $\tilde{x} \in O\tilde{A}$ such that this mapping from OA into $O\tilde{A}$ be linear. When A runs on the boundary of K, we obtain a one-to-one mapping from K onto Σ. The reader can carry out the proof himself.

Corollary   If K is a compact convex set belonging to a linear topological Hausdorff space E of finite dimension, then any continuous mapping of K into itself has at least one fixed point.

Lemma 2.4   Assume E is a locally convex Hausdorff space. Let A be a compact subset of E, and K a convex subset such that $A \subset K$. Then, given an arbitrary neighborhood U of the zero element $\theta \in E$, there exists a continuous mapping $x \to T_U(x)$ from A into E such that

$$T_U(x) \in L \cap K, \qquad x \in A, \tag{2.1}$$

$$T_U(x) - x \in U, \qquad x \in A, \tag{2.2}$$

where L is a finite-dimensional subspace of the space E.

Proof   Without loss of generality we may assume that U is convex and balanced (see, for instance, Yosida [1]). Denote by $|x|_U$ the Minkowski functional associated with U. Then $x \to |x|_U$ is a continuous seminorm on E. Moreover,

$$U = \{x : x \in E, \quad |x|_U < 1\}. \tag{2.3}$$

Since A is compact, there exists a finite set $\{e_k : k = 1, \dots, n\} \subset A$ such that

$$A \subset \bigcup_{i=1}^{n} U(e_i), \tag{2.4}$$

where $U(a) = U + a,\ a \in E$.

Let us consider the following functions:

$$\mu_j(x) = \max\{0, 1 - |x - e_j|_U\}, \qquad x \in E, \quad j = 1, 2, \dots, n. \tag{2.5}$$

Since $|x|_U$ is a continuous function on E, there results that each $\mu_j(x)$, $j = 1, 2, \dots, n$, is a continuous function on E. We have obviously

$$0 \le \mu_j(x) \le 1, \qquad x \in E, \quad j = 1, 2, \dots, n, \tag{2.6}$$

and $\mu_j(x) = 0$ for $x \notin U(e_j)$, $\mu_j(x) > 0$ otherwise. The last statement results from (2.3) and (2.5).

Let us define now

$$T_U(x) = \sum_{i=1}^{n} \mu_i(x)e_i \bigg/ \sum_{i=1}^{n} \mu_i(x), \qquad x \in A. \tag{2.7}$$

From $x \in A$, it follows that $x \in U(e_j)$ for at least one value of j. Hence $\mu_j(x) > 0$, and, taking into account (2.6), one obtains that the denominator in (2.7) is always positive. Consequently, $T_U(x)$ is a continuous function on A. Its values obviously belong to the linear subspace L, generated by $\{e_k : k = 1, 2, \ldots, n\}$. Since $A \subset K$ and K is convex, it follows that the values of $T_U(x)$ also belong to K. This proves (2.1). In order to prove (2.2), we shall remark that

$$T_U(x) - x = \sum_{i=1}^{n} \mu_i(x)(e_i - x) \bigg/ \sum_{i=1}^{n} \mu_i(x). \tag{2.8}$$

From (2.8) we obtain

$$|T_U(x) - x|_U \le \sum_{i=1}^{n} \mu_i(x)|e_i - x|_U \bigg/ \sum_{i=1}^{n} \mu_i(x) < 1, \tag{2.9}$$

because, for any i, either $\mu_i(x) = 0$ and $|e_i - x|_U \ge 1$ or $\mu_i(x) > 0$ and $|e_i - x|_U < 1$. Inequality (2.9) is equivalent to (2.2).

Theorem 2.1 (Schauder–Tychonoff) Assume E is a locally convex Hausdorff space and let $x \to f(x)$ be a continuous mapping from a convex subset $K \subset E$ into E such that

$$f(K) \subset A \subset K, \tag{2.10}$$

with A compact. Then there exists at least one fixed point for f.

Proof Let U be a convex balanced neighborhood of the zero element $\theta \in E$, and consider the function $T_U(x)$, whose existence was shown in the preceding lemma. Define now

$$f_U(x) = T_U(f(x)), \qquad x \in K. \tag{2.11}$$

Since $T_U(x)$ takes its values in the space L, we shall restrict our considerations to this space. From Lemma 2.4 it follows that

$$f_U(L \cap K) \subset T_U(A) \subset L \cap K. \tag{2.12}$$

Indeed, $x \in L \cap K$ implies $f(x) \in A$; hence

$$f_U(x) = T_U(f(x)) \subset L \cap K.$$

Denote by K' the convex hull of the compact set $T_U(A)$ in L. Then K' is also compact (see Valentine [1]). Since $K' \subset L \cap K$, one can easily see that

$$f_U(K') \subset K'. \tag{2.13}$$

Indeed, from (2.12) and $T_U(A) \subset K' \subset L \cap K$, (2.13) follows immediately. But K' is a compact convex set in a finite-dimensional Hausdorff space, and f is continuous. Lemma 2.1 applies, and we obtain the existence of at least one $x \in K'$ such that $x = f_U(x)$. In other words, x satisfies

$$x - f(x) \in U, \tag{2.14}$$

because $x = f_U(x)$ is equivalent to $x = T_U(f(x))$ and, by Lemma 2.3, we have $T_U(f(x)) - f(x) \in U$.

Consequently, the following intermediate result is established: to any neighborhood U of θ, there corresponds at least one $x \in K' \subset K$ such that (2.14) be fulfilled.

Assume now that $x \neq f(x)$ for any $x \in K$. Denote by V_x and W_x two neighborhoods of θ with the properties

$$f(K \cap V_x(x)) \subset W_x(f(x)) \tag{2.15}$$

and

$$V_x(x) \cap W_x(f(x)) = \varnothing. \tag{2.16}$$

The existence of such neighborhoods follows from the continuity of the mapping f and from the fact that E is a Hausdorff space. Let U_x be another neighborhood of θ such that

$$2U_x \subset V_x \cap W_x. \tag{2.17}$$

From the compactness of A there results the existence of a finite set $\{a_k : k = 1, 2, \ldots, m\} \subset A$ such that

$$A \subset \bigcup_{i=1}^{m} U_{a_i}(a_i). \tag{2.18}$$

We shall prove now that for any $x \in K$, there exists an index j, $1 \leq j \leq m$, such that

$$x - f(x) \in U_{a_j} \tag{2.19}$$

cannot hold. Since $y = f(x) \in A$, there exists a j for which $y \in U_{a_j}(a_j)$. It then follows that $U_{a_j}(y) \subset V_{a_j}(a_j)$, because $y = y' + a_j$, $y' \in U_{a_j}$, and $z \in U_{a_j}(y)$ means $x = y'' + y = y'' + y' + a$, with $y'' \in U_{a_j}$. Hence $z \in 2U_{a_j} + a_j \subset V_{a_j}(a_j)$. Assume that (2.19) is true. This would imply $x \in U_{a_j}(y)$ and, consequently, $x \in V_{a_j}(a_j)$. From (2.15) one obtains $y = f(x) \in W_{a_j}(f(a_j))$. But $y \in W_{a_j}(f(a_j))$ and (2.16) lead to $y \notin V_{a_j}(a_j)$, which contradicts (2.17). Therefore, (2.19) cannot be true. Choosing U such that

$$U \subset \bigcap_{i=1}^{m} U_{a_j}, \tag{2.20}$$

it follows that $x - f(x) \notin U$ for any $x \in K$. The last conclusion contradicts the fact we established above that for any neighborhood U of θ there exists at least one $x \in K$ for which (2.14) holds. Theorem 2.1 is thus proved.

Corollary Let K be a closed convex subset of a locally convex Hausdorff space E. Assume that $f: K \to K$ is continuous and that $f(K)$ is relatively compact in E. Then f has at least one fixed point in K. Indeed, one can take $A = \overline{f(K)}$, i.e., the closure of $f(K)$ in E. This corollary is particularly useful for applications.

Exercises

1. Let E be a Banach space and assume $K \subset E$ is a closed convex set. If $f: K \to K$ is weakly continuous and $f(K)$ is relatively compact in the weak topology, then there exists at least one fixed point for the mapping f.

2. (Contraction mapping principle) Let E be a Banach space and $f: E \to E$ be a mapping such that

$$\|f(x) - f(y)\| \le \alpha \|x - y\|, \tag{†}$$

with $0 \le \alpha < 1$. Then there exists a unique fixed point for f, say $x^* \in E$. Show that $x^* = \lim x_n$ as $n \to \infty$, where x_0 is arbitrary in E and $x_n = f(x_{n-1})$, $n \ge 1$.

3. Let E be a Banach space and consider the ball

$$\Sigma = \{x : x \in E, \|x\| \le \rho, \rho < 0\}.$$

Assume $f: \Sigma \to E$ is a contraction mapping (i.e., it satisfies condition (†) in Exercise 2) such that $\|f(\theta)\| \le \rho(1 - \alpha)$, where θ is the zero element of E. Show that $f(\Sigma) \subset \Sigma$. Hence, there exists a unique fixed point $x^* \in \Sigma$.

4. (Krasnoselski) Let K be a convex closed subset of a Banach space E. Assume f and g are mappings from K into E such that the following conditions are satisfied:

a. $f(x) + g(y) \in K$ for $x, y \in K$;
b. f is a contraction mapping, as defined in Exercise 2;
c. g is continuous and carries any bounded set into a relatively compact set.

Then, there exists at least one $x^* \in K$ such that $f(x^*) + g(x^*) = x^*$. In other words, the mapping $h = f + g$ has at least one fixed point in K.

5. Let E be a Banach space and consider a linear operator $A: E \to E$. Assume that $f: E \to E$ is a mapping satisfying a Lipschitz condition. Discuss the existence of solutions of the equation $x = Ax + f(x)$.

1.3   Fourier and Laplace Transforms

In view of their frequent use in the subsequent chapters, we shall deal briefly with Fourier and Laplace transforms. Our main goal is to formulate one of the most interesting properties of these transforms, which—roughly speaking—states that the class of Fourier (Laplace) transforms is closed with respect to analytic operations. The theory of normed rings is particularly adequate in order to express such properties (see Gel'fand *et al.* [1], Yosida [1]).

Assume E is a commutative normed ring over the complex number field with a unit element e such that $\|e\| = 1$. By a commutative normed ring E we understand a Banach space E over the complex number field such that an inner multiplication is given on E. Moreover, we have $\|xy\| \leq \|x\| \|y\|$.

For instance, the set $C(S)$ of all continuous complex-valued functions on a compact topological space S becomes a commutative normed ring if we define

$$(x + y)(s) = x(s) + y(s), \qquad (xy)(s) = x(s)y(s), \qquad (\alpha x)(s) = \alpha x(s)$$

for any $x, y \in C(S)$, $s \in S$, and complex number α. The norm is given by

$$\|x\| = \sup\{|x(s)| : s \in S\}.$$

The ideal theory of normed rings plays an important role in the investigation of many fundamental problems related to this field.

We recall that a subset A of E is an ideal if and only if $x, y \in A$ implies $\alpha x + \beta y \in A$ for any complex α, β, and $x \in A$, $z \in E$ imply $xz \in A$. An ideal that does not coincide with $\{\theta\}$ or E is called a proper ideal or a nontrivial ideal.

A nontrivial ideal M of E is called a maximal ideal if and only if from $M \subset A$, where A denotes an ideal of E, there results $M = A$ or $A = E$. In other words, there exists no proper ideal of E containing M as a proper subset.

The following basic result can be found in the books we mentioned above.

Theorem 3.1   The quotient algebra E/M, where M is a maximal ideal of E, is isometrically isomorphic to the complex number field.

In other words, Theorem 3.1 states that $E/M = \mathscr{C}\bar{e}$, where $\mathscr{C}$ denotes the complex number field.

Remark 1   If $X \in E/M$, then $\|X\| = \inf\{\|y\| : y \in X\}$.

Remark 2   The isomorphism whose existence is stated in Theorem 2.1 is called the canonical isomorphism.

Consequently, to any $X \in E/M$, we can uniquely associate a complex number ξ_X. If $x \in X$, then we define $x(M) = \xi_X$. When M runs over the set of all maximal ideals of E, we get a complex-valued function $M \to x(M)$.

Among the properties of the mapping $M \to x(M)$, associated to the element $x \in E$ and defined on the set $\{M\}$ of all maximal ideals of E, we shall indicate the following one:

Theorem 3.2   The spectrum of any element $x \in E$ coincides with the range of the associated mapping $M \to x(M)$.

If $x \in E$, then the spectrum of x consists of all complex λ's such that $(\lambda e - x)^{-1}$ does not exist. The spectrum is always a nonempty compact set.

We have $|x(M)| \leq \|x\|$, $M \in \{M\}$. Indeed, if $X \in E/M$ is the residue class containing x, then $\|X\| = \inf\{\|y\| : y \in X\}$ by definition. Hence $\|X\| \leq \|x\|$. Let ξ_X be the complex number that corresponds to X in the canonical isomorphism of E/M on the complex number field. Then $\|X\| = |\xi_X| = |x(M)|$ and the inequality is proved.

Before proceeding further, we shall remark that a theory of analytic functions of a complex variable, with values in a commutative normed ring E, can be easily built up following the classical model of analytic complex-valued functions. Such concepts and results as Cauchy's theorem, Cauchy's integral formula, series expansions, and many others remain valid without essential changes with respect to the classical model.

We are now able to state the main result of this section.

Theorem 3.3   Let E be a commutative normed ring with unit e, over the field of complex numbers. If $x \in E$ and $f(\lambda)$ is an analytic function in the neighborhood of the spectrum of x, with values in E, then

$$y = f(x) = (2\pi i)^{-1} \int_\Gamma (\lambda e - x)^{-1} f(\lambda) \, d\lambda, \tag{3.1}$$

where Γ is a contour that contains the spectrum of x and belongs to the domain of analiticity for $f(\lambda)$, defines an element $y \in E$ such that

$$y(M) = f(x(M)) \tag{3.2}$$

for any maximal ideal M of E.

Proof   Let us remark first that $g(\lambda) = (\lambda e - x)^{-1} f(\lambda)$ is an analytic (holomorphic) function of λ in a neighborhood of the contour Γ. Hence, the integral in the right-hand side of (3.1) exists and is independent of the choice of Γ, provided we are dealing with contours that satisfy the conditions in the theorem. The independence follows from Cauchy's theorem for analytic functions with values in E. Now, let us fix the maximal ideal M and consider the mapping $x \to x(M) = M(x)$ from E into the complex number field. This mapping is linear and continuous. As seen above, $|M(x)| \le \|x\|$. This allows us to write

$$y(M) = (2\pi i)^{-1} \int_\Gamma (\lambda e - x)^{-1}(M) f(\lambda)\, d\lambda$$

$$= (2\pi i)^{-1} \int_\Gamma [f(\lambda)/(\lambda - x(M))]\, d\lambda = f(x(M)).$$

The theorem is thus proved.

Remark   The following relationship was needed above:

$$(\lambda e - x)^{-1}(M) = (\lambda - x(M))^{-1}.$$

Since $(\lambda e - x)(M) = \lambda - x(M)$, and from $x = yz$ we have $x(M) = y(M)z(M)$, it suffices to observe that $e(M) = 1$ (the unit of the algebra E/M is carried onto the unit of the complex number field by the canonical isomorphism).

Many interesting results can be derived from Theorem 3.3. We shall consider below two applications related to Fourier and Laplace transforms.

Let us consider the space $L = L(R, \mathscr{C})$ of complex-valued functions such that $x \in L$ if and only if $|x| \in L(R, R)$. If we define the convolution product by

$$(x * y)(t) = \int_R x(t - s)y(s)\, ds \tag{3.3}$$

for any $x, y \in L$, then L becomes a commutative normed ring (with the usual norm of L). It can be shown that L does not contain a unit element (see Gel'fand *et al.* [1, Chapter III, §16]).

One can easily extend L to a commutative normed ring with a unit element. Indeed, consider the set of all pairs (λ, x), with $\lambda \in \mathscr{C} =$ complex number field and $x \in L$. The sum, the scalar multiplication, and the product are defined, respectively, by

$$(\alpha, x) + (\mu, y) = (\lambda + \mu, x + y),$$
$$\alpha(\lambda, x) = (\alpha\lambda, \alpha x),$$
$$(\lambda, x) * (\mu, y) = (\lambda\mu, \lambda y + \mu x + x * y).$$

The norm is $\|(\lambda, x)\| = |\lambda| + \|x\|_L$. Now, it is easy to see that $V = \{(\lambda, x) : \lambda \in \mathscr{C}, x \in L\}$ is a commutative normed ring with unit element $e = (1, 0)$. Since $(\lambda, x) = (\lambda, 0) + (0, x)$, one can write any element of V in the form $\lambda e + x$, if we agree to identify the pair $(0, x)$ with $x \in L$.

Obviously, L is a maximal ideal of V. It is denoted by M_∞. The meaning of this notation will become clear below.

The following theorem can be also found in Gel'fand *et al.* [1, Chapter III, §17] and it clarifies completely the structure of the set of maximal ideals of V.

Theorem 3.4 Let M be a maximal ideal of V such that $M \neq M_\infty$. Then there exists a real number s with the property that for any $\lambda e + x \in V$,

$$(\lambda e + x)(M) = \lambda + \int_R x(t)e^{ist} \, dt. \tag{3.4}$$

Now, it is clear that the Fourier transform appears in a natural manner in connection with the ideal theory of commutative normed rings. As we know,

$$\tilde{x}(s) = \int_R x(t)e^{ist} \, dt, \qquad s \in R, \tag{3.5}$$

is nothing but the *Fourier transform* of the function $x \in L(R, \mathscr{C})$.

Theorem 3.5 (of Wiener and Lévy, for Fourier transforms) Let $x \in L(R, \mathscr{C})$ be a function and $\tilde{x}(s)$ be its Fourier transform. Assume $\phi(z)$ is an analytic complex-valued function in a neighborhood of the curve $z = \tilde{x}(s)$, $s \in R$. If $\phi(0) = 0$, then there exists $y \in L(R, \mathscr{C})$ such that

$$\tilde{y}(s) = \phi(\tilde{x}(s)), \qquad s \in R. \tag{3.6}$$

Proof Consider the commutative normed ring consisting of all functions of the form $\lambda + \tilde{x}(s)$, where $\tilde{x}(s)$ is the Fourier transform of a function $x \in L(R, \mathscr{C})$ and λ is an arbitrary complex number. That this class of functions forms a ring with respect to the usual operations of addition and multiplication results easily from the well-known property of Fourier transforms:

$$\widetilde{x * y} = \tilde{x}\tilde{y}. \tag{3.7}$$

The norm is $\|\lambda + \tilde{x}\| = |\lambda| + \|x\|_L$, i.e., we are using the same norm as in V. In this ring, the functions that are the Fourier transforms of the functions from $L(R, \mathscr{C})$ are characterized by the property that they tend to zero as $|s| \to \infty$. In other words, they correspond to $\lambda = 0$ (because $\tilde{x}(s) \to 0$ as $|s| \to \infty$, for any $x \in L$). Since zero belongs to any neighborhood of the curve $z = \tilde{x}(s)$, $s \in R$, the condition $\phi(0) = 0$ from the statement always has a meaning. From Theorem 3.3, it follows that $\phi(\tilde{x}(s))$ is of the form $\lambda + \tilde{y}(s)$, with $y \in L(R, \mathscr{C})$. Now letting $|s| \to \infty$ in $\phi(\tilde{x}(s)) = \lambda + \tilde{y}(s)$, we obtain $\lambda = 0$. Hence (3.6) holds.   Q.E.D.

Consider now the space $L(R_+, \mathscr{C})$. If $x \in L(R_+, \mathscr{C})$, then

$$x_1(t) = \begin{cases} x(t), & t \in R_+, \\ 0, & t < 0, \end{cases} \tag{3.8}$$

belongs to $L(R, \mathscr{C})$. In other words, the space $L(R_+, \mathscr{C})$ can be isomorphically imbedded into $L(R, \mathscr{C})$. If $x, y \in L(R_+, \mathscr{C})$, then formula (3.3) becomes

$$(x * y)(t) = \int_0^t x(t - s)y(s) \, ds, \qquad (3.9)$$

i.e., we obtain the convolution product usually related to the one-sided transform. Formula (3.5) becomes

$$\tilde{x}(s) = \int_0^\infty x(t)e^{ist} \, dt \qquad (3.10)$$

and $\tilde{x}(s)$ is now defined for any complex s such that $\operatorname{Im} s \geq 0$. In other words the Laplace transform of a function $x \in L(R_+, \mathscr{C})$ is an analytic function in the half-plane $\operatorname{Im} s > 0$, continuous in the closed half-plane.

It is easy to see that the set of all elements of V which are of the form $\lambda e + x$, with $x \in L(R_+, \mathscr{C})$ (we agree to identify $L(R_+, \mathscr{C})$ with its isomorphic image in $L(R, \mathscr{C})$), forms a subring of V. It will be denoted by V_+. It turns out (see Gel'fand *et al.* [1, Chapter III, §16]) that, excepting the maximal ideal consisting of all elements of the form $(0, x)$, $x \in L(R_+, \mathscr{C})$, any other maximal ideal M of V_+ is determined by a complex number s with $\operatorname{Im} s \geq 0$. Corresponding to (3.4) in the case of the ring V, we have

$$(\lambda e + x)(M) = \lambda + \int_0^\infty x(t)e^{ist} \, dt, \qquad (3.11)$$

where s belong to the half-plane $\operatorname{Im} s \geq 0$.

Now, we can formulate a theorem that is similar to Theorem 3.5.

Theorem 3.6 (of Wiener and Lévy, for the Laplace transform) Let $x \in L(R_+, \mathscr{C})$ be a function whose Laplace transform $\tilde{x}(s)$ is given by (3.10). Assume $\phi(z)$ is an analytic complex-valued function in a neighborhood of the range of $\tilde{x}(s)$, $\operatorname{Im} s \geq 0$, such that $\phi(0) = 0$. Then there exists $y \in L(R_+, \mathscr{C})$ such that

$$\tilde{y}(s) = \phi(\tilde{x}(s)), \qquad \operatorname{Im} s \geq 0. \qquad (3.12)$$

The task of carrying out the proof of Theorem 3.6 is left to the reader.

Exercises

1. The following properties hold for the Fourier transform:

a. The transform of the convolution product $x * y$, with $x, y \in L(R, \mathscr{C})$, is the (ordinary) product $\tilde{x}\tilde{y}$ of the transforms (see formula (3.7)).

b. If $x \in L(R, \mathscr{C}) \cap L^2(R, \mathscr{C})$, then $\tilde{x} \in L^2(R, \mathscr{C})$ and Parseval's formula holds:

$$\int_R |\tilde{x}(s)|^2 \, ds = 2\pi \int_R |x(t)|^2 \, dt.$$

c. If $x, y \in L(R, \mathscr{C}) \cap L^2(R, \mathscr{C})$, then

$$\int_R (\tilde{x}\bar{\tilde{y}} + \bar{\tilde{x}}\tilde{y}) \, ds = 2\pi \int_R (x\bar{y} + \bar{x}y) \, dt.$$

2. Show that for any $x \in L(R, \mathscr{C})$, $\tilde{x} \in C_0(R, \mathscr{C})$. Find examples such that $\tilde{x} \notin L(R, \mathscr{C})$.

3. Assume that $x \in L(R, \mathscr{C})$ is such that $\tilde{x} \in L(R, \mathscr{C})$. Then

$$x(t) = (2\pi)^{-1} \int_R \tilde{x}(s) e^{-ist} \, ds, \qquad \text{a.e. on } R.$$

This is an inversion theorem for Fourier transform.

4. An L^2 theory of Fourier transform can be built up as follows. First, observe that $L^1 \cap L^2$ is dense in L^2. Since the Fourier transform is well defined for any $x \in L^1 \cap L^2$ and $\tilde{x} \in L^2$ (see Exercise 1 above), we can extend the mapping $x \to \tilde{x}$ to L^2 by continuity. This extension is unique and presents more symmetry because both x and $\tilde{x}$ belong to L^2. Show that

$$\tilde{x}(s) = \lim_{T \to \infty} \int_{|t| \le T} x(t) e^{ist} \, dt,$$

the limit being that of the L^2 topology.

5. Let E be a commutative normed ring and assume M is a maximal ideal of E. Show that M is a closed subset of E.

6. Let $f: E \to \mathscr{C}$ be a multiplicative linear functional on the commutative normed ring E, with values in the complex number field $\mathscr{C}$. This means that f is a linear functional on the Banach space E such that $f(x\,y) = f(x)f(y)$, for any $x, y \in E$. Show that the set of all $x \in E$ such that $f(x) = 0$ is a maximal ideal of E.

7. Let E be the class of continuous complex-valued functions on $[-\pi, \pi]$, which can be represented as an absolutely convergent Fourier series

$$x(t) = \sum_{n=-\infty}^{\infty} c_n e^{int},$$

with $\sum_{-\infty}^{\infty} |c_n| < +\infty$. Show that E can be organized as a commutative normed ring, the norm being given by $\|x\| = \sum_{n=-\infty}^{\infty} |c_n|$.

8. Show that the set of all linear continuous operators of a Banach space can be organized as a normed ring (generally, noncommutative).

9. Let $M = M(R_+, \mathscr{C})$ be the space of complex-valued locally integrable functions on R_+ such that $x \in M$ if and only if

$$\sup\left\{\int_t^{t+1} |x(s)| \; ds : t \in R_+\right\} < +\infty.$$

Show that the Laplace transform $\tilde{x}(s)$ as given by (3.10) is defined for $\operatorname{Im} s > 0$.

1.4   A Factorization Problem

The reader is undoubtedly aware of several factorization problems encountered in classical analysis. The aim of a factorization is to represent a given function as a product (finite or infinite) of simpler functions. For instance, any entire function $f(z)$ of a complex variable can be represented as a product (generally, infinite) of the form

$$f(z) = \exp\{g(z)\} \prod_{n=1}^{\infty} E_{p_n}(z/z_n),$$

where $g(z)$ is another entire function, $\{z_n : n = 1, 2, \ldots\}$ is the sequence of the zeros of $f(z)$, and

$$E_p(z) = (1 - z) \exp\{z + z^2/2 + \cdots + z^p/p\},$$

$\{p_n : n = 1, 2, \ldots\}$ is a convenient sequence of nonnegative integers. This is the famous Weierstrass factorization theorem.

The factorization problem we are going to discuss here is somewhat different from that mentioned above and appears in connection with the so called Wiener–Hopf technique in the theory of integral equations.

Consider a function $F \in C_\ell(R, \mathscr{C})$, i.e., a continuous complex-valued function defined on the real axis R such that both limits at $\pm \infty$ exist. These limits will be denoted by $F(\infty)$ and $F(-\infty)$, respectively. If they are equal, it is obvious that F can be considered a continuous function on the closed line $\bar{R}$.

Assume F can be represented in the form

$$F(s) = F_+(s)F_-(s), \qquad s \in \bar{R}, \tag{4.1}$$

where $F_+(s)$ is holomorphic in the half-plane $\operatorname{Im} s > 0$ and continuous in the closed half-plane $\operatorname{Im} s \geq 0$, while $F_-(s)$ is holomorphic for $\operatorname{Im} s < 0$ and continuous for $\operatorname{Im} s \leq 0$. This is the kind of factorization with which we shall be concerned.

More specifically, we shall consider only such $F \in C_\ell(R, \mathscr{C})$ that can be represented as

$$F(s) = 1 - G(s), \qquad s \in R, \tag{4.2}$$

where $G(s)$ is the Fourier transform of a certain function $g \in L(R, \mathscr{C})$:

$$G(s) = \int_R g(t)e^{ist}\, dt, \qquad s \in R. \tag{4.3}$$

We know that $G(\infty) = G(-\infty) = 0$, whence $F(\infty) = F(-\infty) = 1$. One agrees to consider only such factorizations (4.1), with $F_+(\infty) = F_-(\infty) = 1$.

We shall assume throughout this section that $F(s)$ is such that

$$F(s) \neq 0, \qquad s \in R. \tag{4.4}$$

This condition implies, of course, that $F_+(s)$ and $F_-(s)$ cannot vanish on R.

The factorization (4.1) is called canonical if both $F_+(s)$ and $F_-(s)$ do not vanish for $\operatorname{Im} s \geq 0$ and $\operatorname{Im} s \leq 0$, respectively.

Before proceeding further, we shall recall first the definition and some properties of the concept of index of a closed path in the complex plane.

Consider the path whose equation is $z = F(s)$, $s \in \overline{R}$, where $F(s)$ is given by (4.2) and satisfies (4.4). Then

$$\operatorname{ind} F = (2\pi)^{-1}[\arg F(s)]_{s=-\infty}^{s=\infty} \tag{4.5}$$

is an integer. It is, by definition, the index of the closed path or—equivalently —the index of the function $F(s)$.

A very simple but useful property, following directly from the definition, is expressed by the following formula:

$$\operatorname{ind}(F_1 F_2) = \operatorname{ind} F_1 + \operatorname{ind} F_2. \tag{4.6}$$

This means that both F_1 and F_2 satisfy the conditions we required above for $F(s)$. A particular case of (4.6) corresponds to $F_2 = \alpha = \text{constant}$. This yields $\operatorname{ind}(\alpha F) = \operatorname{ind} F$.

A basic result concerning our factorization problem is the following.

Theorem 4.1 Consider a function $F(s)$ of the form (4.2), with $G(s)$ given by (4.3). A necessary and sufficient condition for the existence of a canonical factorization of the form (4.1) is that (4.4) and

$$\operatorname{ind} F = 0 \tag{4.7}$$

hold. The canonical factorization is then unique. Moreover, we have

$$F_+(s) = 1 + \int_0^\infty \gamma(t)e^{ist}\, dt, \qquad \operatorname{Im} s \geq 0 \tag{4.8}$$

$$F_-(s) = 1 + \int_0^\infty \mu(t)e^{-ist}\, dt, \qquad \operatorname{Im} s \leq 0 \tag{4.9}$$

with $\gamma, \mu \in L(R_+, \mathscr{C})$.

Proof It is obvious that (4.4) is necessary. In order to prove that (4.7) is also necessary, we shall remark first that ind $F = $ ind $F_+ + $ ind F_-. Now, it can be easily seen that both ind F_+ and ind F_- are zero. Indeed, on the Riemann sphere obtained by the compactification of the complex plane, we have a circumference that corresponds to the real line such that F_+ is analytic on a hemisphere determined by this circumference and does not vanish. By using the theorem concerning the variation of the argument of an analytic function along the boundary of the analyticity domain, we obtain ind $F_+ = 0$.

Similar arguments hold in the case of F_-. Therefore, ind $F = 0$, and this means that (4.7) is necessary for the existence of a canonical factorization.

Under conditions (4.4) and (4.7) one can prove, using Theorem 3.5, that there exists $h \in L(R, \mathscr{C})$ such that

$$F(s) = \exp\left\{\int_R h(t)e^{ist}\,dt\right\}, \qquad s \in R. \tag{4.10}$$

Let us define now

$$F_+(s) = \exp\left\{\int_0^\infty h(t)e^{ist}\,dt\right\}, \qquad \text{Im } s \geq 0, \tag{4.11}$$

and

$$F_-(s) = \exp\left\{\int_{-\infty}^0 h(t)e^{ist}\,dt\right\}, \qquad \text{Im } s \leq 0. \tag{4.12}$$

It is clear that (4.1) holds and that this factorization is canonical.

The uniqueness of the canonical factorization can be established as follows. Assume $F(s) = F_+^*(s)F_-^*(s)$ is another canonical factorization of F. Then

$$F_+(s)/F_+^*(s) = F_-^*(s)/F_-(s), \qquad s \in R, \tag{4.13}$$

The right-hand side of (4.13) is an analytic function in the half-plane Im $s > 0$, continuous in the closed half-plane, while the left-hand side is analytic for Im $s < 0$ and continuous in the closed half-plane.

Since both sides coincide on the real axis, it follows that they are restrictions of an analytic function in the whole complex plane to the half-planes Im $s \geq 0$ and Im $s \leq 0$, respectively. Liouville's theorem applies and we obtain that both sides in (4.13) equal the same constant. It can be easily seen, taking into account the values at infinity, that this constant is 1. Consequently, the canonical factorization is unique.

In order to obtain (4.8) and (4.9) from (4.11) and (4.12), respectively, we have to apply Theorem 3.6. The factorization problem in the case ind $F \neq 0$

can also be discussed using arguments similar to those encountered above. A slight modification of the properties required for $F_+(s)$ and $F_-(s)$ is necessary.

We shall say that the factorization (4.1) is regular if at least one of the factors does not vanish in the corresponding half-plane of analyticity.

Theorem 4.2   Let $F(s)$ be a function of the form (4.2), where $G(s)$ is given by (4.3). Assume that (4.4) holds and

$$v = \text{ind } F > 0. \tag{4.14}$$

Let $s_1, s_2, \ldots, s_m$, $m \leq v$, be some arbitrary points in the half-plane Im $s > 0$ and consider m positive integers $p_1, p_2, \ldots, p_m$, with $p_1 + p_2 + \cdots + p_m = v$. Then there exists a unique regular factorization of the form (4.1), such that $F_+(s)$ has the points $s_1, s_2, \ldots, s_m$ as zeros of multiplicities $p_1, p_2, \ldots, p_m$, respectively, and has no other zeros. Formulas similar to (4.8) and (4.9) hold for $F_+(s)$ and $F_-(s)$, respectively.

Proof   Let us remark first that

$$R(s) = \prod_{k=1}^{m} \left[\frac{(s - s_k)}{(s + i)} \right]^{p_k} \tag{4.15}$$

satisfies ind $R(s) = v$. (This follows from the theorem concerning the variation of the argument of an analytic function along the boundary of the analyticity domain, taking into account that $R(s)$ has v zeros in the half-plane Im $s > 0$.)

Since for real s, $1/R(s)$ is of the form $1 +$ the Fourier transform of a certain function (which one?), it follows that $F(s)/R(s)$ is again of the form (4.2). Moreover, we have $\text{ind}[F(s)/R(s)] = 0$, which allows us to apply Theorem 4.1. Hence, $F(s)/R(s) = F_+{}^*(s)F_-{}^*(s)$, the factorization being canonical. If we denote $F_+(s) = F_+{}^*(s)R(s)$, $F_-(s) = F_-{}^*(s)$, then $F(s) = F_+(s)F_-(s)$ and we obtained the desired regular factorization.

Remark   If condition (4.14) is replaced by

$$v = \text{ind } F < 0, \tag{4.16}$$

then a similar statement holds, changing v to $|v|$ and referring to $F_-(s)$ instead of $F_+(s)$.

In concluding this section, we shall remark that similar results to those given in Theorems 4.1 and 4.2 can be obtained for nonscalar $F(s)$. The case when $F(s)$ is a matrix function was discussed, using complementary arguments, by Gochberg and Krein [1].

Exercises

1. Assume that (4.1) is the canonical factorization for $F(s)$. Show that

$$\ln F_+(s) = (2\pi i)^{-1} \int_R [\ln F(u)/(u - s)] \, du, \qquad \text{Im } s > 0$$

$$\ln F_-(s) = -(2\pi i)^{-1} \int_R [\ln F(u)/(u - s)] \, du, \qquad \text{Im } s < 0.$$

2. Assume that $k \in L(R, \mathscr{C})$ and

$$\tilde{k}(s) = \int_R k(t) e^{ist} \, dt$$

is a rational function such that

$$\tilde{k}(s) = 1 - \prod_{j=1}^{n} [(s - s_j)/(s - \sigma_j)],$$

with $\text{Im } s_j < 0$, $\text{Im } \sigma_j < 0$, for $j = 1, 2, \ldots, m < n$, and $\text{Im } s_j > 0$, $\text{Im } \sigma_j > 0$ for $j = m + 1, \ldots, n$. If $F(s) = [1 - \tilde{k}(s)]^{-1}$, $s \in R$, show that ind $F = 0$ and find the canonical factorization of $F(s)$.

2

Admissibility and Hammerstein Equations

We shall be concerned in this chapter with the investigation of some classes of nonlinear integral equations that are usually called Hammerstein equations. The main feature that distinguishes the subsequent results from the classical ones is the fact that the domain of definition of the functions involved in our considerations has infinite measure (a half-axis or the entire real axis).

Since linear integral operators are essentially related to this kind of equation, we shall first consider several problems concerning these operators. The most important problem for our further developments is that of continuity. Nevertheless, several additional properties presenting interest for applications will be discussed.

It is useful to point out that the results we shall present for linear integral operators are nothing but stability results concerning linear systems for which the input–output equation is given by means of an integral transform.

The last part of this chapter contains various existence theorems for Hammerstein integral equations. We shall frequently use fixed-point theorems and the results established in the first part of the chapter concerning linear integral operators. The fact that the solution belongs to a certain function space allows us to describe its properties more easily (for instance, the asymptotic behavior at infinity).

2.1 Admissibility and Continuity

Consider the linear integral operator

$$(Kx)(t) = \int_R k(t, s)x(s) \, ds, \qquad t \in R, \tag{1.1}$$

where x belongs to a certain function space $E = E(R, R^n)$ and $k(t, s)$ is a matrix function of type m by n, whose elements are at least measurable.

Assume that for any $x \in E$ we have $Kx \in F$, where $F = F(R, R^m)$ is also a function space. When this property holds, we shall say that the pair of function spaces (E, F) is *admissible* with respect to the operator K.

Let us remark that no requirement was made on the continuity of the operator K in defining the admissibility. Actually, the continuity of K follows almost automatically from the admissibility property or from other properties pertaining to the special structure of this operator (for instance, from the property of being closed). Thus, when E and F are Banach spaces or even Fréchet spaces, the closed-graph theorem applies without difficulty, and this yields the continuity of the integral operator K.

Let us recall some definitions and results related to the concept of Fréchet space. As a basic reference, see Dunford and Schwartz [1, Chapter II, §§ 1, 2].

A linear space E is called a Fréchet space if a distance function ρ is given on E such that the following conditions are verified: (a) the distance is invariant, i.e., $\rho(x, y) = \rho(x - y, \theta)$ for any $x, y \in E$; (b) the map $(\lambda, x) \to \lambda x$ is continuous with respect to each argument; (c) the space is complete in the topology derived from the metric ρ.

Let E and F be two Fréchet spaces and assume that K is a linear map of E into F. The set $\{(x, Kx) : x \in E\} \subset E \times F$ is called the graph of K. It is obviously a linear manifold in the product space $E \times F$.

The closed graph theorem states that a linear map of a Fréchet space E into a Fréchet space F is continuous if and only if its graph is closed.

Another tool for establishing the continuity and for finding necessary and sufficient conditions under which this property holds for integral operators will be the uniform boundedness principle (called also the Banach–Steinhaus theorem; see, for instance, Rudin [1, Section 5.8]).

A very frequent situation that occurs in the investigation of continuity of linear operators can be described by means of the following scheme.

Lemma 1.1 Let E and F be two linear topological Hausdorff spaces and assume $K: E \to F$ is a continuous linear mapping. Assume further that $E_1 \subset E$ and $F_1 \subset F$ are Fréchet spaces whose topologies are stronger than the topologies of E and F, respectively. If $KE_1 \subset F_1$, then K is continuous from E_1 to F_1.

Proof Let $\{(x_n, Kx_n)\} \subset E_1 \times F_1$ be a convergent sequence of points belonging to the graph of K. This means that $x_n \to x$ in E_1 and $Kx_n \to y$ in F_1. We shall prove that (x, y) belongs to the graph of K, i.e., $y = Kx$. Indeed, $x_n \to x$ in E_1 implies $x_n \to x$ in E, and hence $Kx_n \to Kx$ in F. But $Kx_n \to y$ in F_1 implies $Kx_n \to y$ in F. Therefore, $Kx = y$ because F is a Hausdorff space.

An application of Lemma 1.1 will be given in the next section.

2.2 Admissibility of the Pair (C_g, C_G) with Respect to the Volterra Operator

Assume that the kernel $k(t, s)$ vanishes for $s < 0$ and for $s > t$. Then (1.1) becomes

$$(Kx)(t) = \int_0^t k(t, s)x(s)\, ds. \tag{2.1}$$

We shall discuss in this section the problem of admissibility of the pair (C_g, C_G), $C_g = C_g(R_+, R^n)$, $C_G = C_G(R_+, R^m)$, with respect to the operator K given by (2.1). A necessary and sufficient condition will be obtained for the kernel $k(t, s)$ in order to assure this property of admissibility. First, let us prove the following.

Lemma 2.1 Assume that $k(t, s)$ is a continuous matrix function for $0 \le s \le t < +\infty$, of type m by n. Then the integral operator K, defined by (2.1), is continuous from $C_c(R_+, R^n)$ to $C_c(R_+, R^m)$.

Proof Let $\{x^p\} \subset C_c(R_+, R^n)$ be a convergent sequence to x. This means that $x^p(t) \to x(t)$ as $p \to \infty$, uniformly on any compact interval $[0, T] \subset R_+$. Since

$$(Kx^p)(t) - (Kx)(t) = \int_0^t k(t, s)[x^p(s) - x(s)]\, ds$$

and $k(t, s)$ is bounded on any set of the form $0 \le s \le t \le T$, $T < +\infty$, it follows that $Kx^p \to Kx$ in $C_c(R_+, R^m)$ as $p \to \infty$.

Now, we are able to prove the following result.

Theorem 2.1 Let $k(t, s)$ be a continuous matrix function for $0 \leq s \leq t < +\infty$, of type m by n. Then, a necessary and sufficient condition for the admissibility of the pair (C_g, C_G) with respect to the operator K given by (2.1) is

$$\int_0^t \|k(t, s)\| g(s) \, ds \in C_G(R_+, R). \tag{2.2}$$

Proof One can easily check that (2.2) is a sufficient condition for admissibility. Indeed, condition (2.2) means that there exists a positive constant A such that

$$\int_0^t \|k(t, s)\| g(s) \, ds \leq A G(t), \qquad t \in R_+. \tag{2.3}$$

If $x \in C_g(R_+, R^n)$, then

$$\left\| \int_0^t k(t, s) x(s) \, ds \right\| \leq \int_0^t \|k(t, s)\| g(s)(\|x(s)\| / g(s)) \, ds$$

$$\leq A G(t) |x|_{C_g}, \qquad t \in R_+,$$

which shows that $Kx \in C_G$.

In order to prove the necessity of condition (2.2), we consider first the particular case $m = n = 1$. In other words, we are dealing with scalar functions.

Assume now that condition (2.2) does not hold and prove that

$$K C_g \subset C_G \tag{2.4}$$

is impossible. Indeed, if (2.2) is not satisfied, there exists a sequence $\{t_m\} \subset R_+$ (it is easy to see that $t_m \to \infty$ as $m \to \infty$) such that

$$\int_0^{t_m} |k(t_m, s)| g(s) \, ds > m G(t_m), \qquad m \geq 1. \tag{2.5}$$

Let us denote

$$\varphi_m(t) = g(t) \operatorname{sign} k(t_m, t), \qquad t \in [0, t_m]. \tag{2.6}$$

It is obvious that $\varphi_m(t)$ is a measurable function on $[0, t_m]$ and $|\varphi_m(t)| \leq g(t)$. From (2.5) we derive

$$\int_0^{t_m} k(t_m, s) \varphi_m(s) \, ds > m G(t_n), \qquad m \geq 1. \tag{2.7}$$

A theorem of Lusin concerning measurable functions applies (see, for instance, Rudin [1, Section 2.23]) and we obtain the following intermediate result. There exists a continuous function on $[0, t_m]$, say $f_m(t)$, such that

$$|f_m(t)| \leq g(t), \qquad t \in [0, t_m], \tag{2.8}$$

and

$$\int_0^{t_m} k(t_m, s) f_m(s) \, ds > m G(t_m), \qquad m \geq 1. \tag{2.9}$$

It is obviously possible to extend the function $f_m(t)$ to the whole R_+ such that (2.8) and continuity are preserved. The sequence $\{f_m(t)\}$ belongs to the unit ball in $C_g(R_+, R)$ centered at the origin. If (2.4) is true, then Lemma 1.1 applies and we find that K is continuous from C_g to C_G. Consequently, $\{Kf_m\}$ should be bounded in $C_G(R_+, R)$. From (2.9) we see that this property cannot hold. Consequently, the hypothesis that (2.2) does not hold implies $KC_g \not\subset C_G$.

Let us consider now the general case when $k(t, s) = (k_{ij}(t, s))$, $i = 1, 2, \ldots, m$, $j = 1, 2, \ldots, n$, is a matrix kernel. It is useful to observe that $x \in C_g(R_+, R^n)$ if and only if each coordinate of x belongs to $C_g(R_+, R)$. This fact follows easily from the definition of the space C_g. Let us now take $x \in C_g(R_+, R^n)$ such that $x_k(t) \equiv 0$ on R_+, $k \neq j$, where j is fixed. In other words, only one coordinate of x is nonzero. Since $Kx \in C_G(R_+, R^m)$, we get

$$\int_0^t k_{ij}(t, s) x_j(s) \, ds \in C_G(R_+, R), \tag{2.10}$$

for any i, $i = 1, 2, \ldots, m$, and fixed j. But x_j is arbitrary in $C_g(R_+, R)$. Therefore, taking into account the particular case we discussed above, one obtains from (2.10)

$$\int_0^t |k_{ij}(t, s)| g(s) \, ds \in C_G(R_+, R), \tag{2.11}$$

$i = 1, 2, \ldots, m$. Since j can be chosen arbitrarily among the numbers $1, 2, \ldots, n$, it follows from (2.11) that

$$\int_0^t \|k(t, s)\| g(s) \, ds \in C_G(R_+, R).$$

Theorem 2.1 is thus proved.

Remark   From Lemma 1.1 and Theorem 2.1 it follows that (2.2) is a necessary and sufficient condition for the continuity of the Volterra operator (2.1) from C_g to C_G.

Corollary   For $G(t) \equiv 1$, one obtains the following necessary and sufficient condition that the pair (C_g, C) be admissible with respect to the operator K given by (2.1):

$$\int_0^t \|k(t, s)\| g(s) \, ds \leq M, \qquad t \in R_+, \tag{2.12}$$

where M is a positive number.

2.3   Admissibility of the Pair (C_g, C_ℓ)

We shall consider in this section the problem of admissibility of the pair (C_g, C_ℓ) with respect to the operator K given by (2.1), where C_ℓ is the subspace of C defined in Section 1.1.

The following hypothesis will be assumed throughout this section:

A: $k(t, s)$ is a continuous matrix function for $0 \leq s \leq t < +\infty$, of type m by n, such that

$$\lim_{t \to \infty} k(t, s) = k(s), \qquad s \in R_+, \tag{3.1}$$

uniformly on any compact interval $[0, s_1] \subset R_+$.

It follows that $k(s)$ is a continuous matrix function on R_+, of type m by n. We can now state the following.

Theorem 3.1   Consider the integral operator K given by (2.1), with $k(t, s)$ satisfying hypothesis A. Then, a necessary and sufficient condition for the admissibility of the pair (C_g, C_ℓ) with respect to K is that both conditions (3.2) and (3.3) be satisfied:

$$\int_0^\infty \|k(s)\| g(s) \, ds < +\infty, \tag{3.2}$$

$$\lim_{t \to \infty} \int_0^t \|k(t, s)\| g(s) \, ds = \int_0^\infty \|k(s)\| g(s) \, ds. \tag{3.3}$$

If (C_g, C_ℓ) is admissible with respect to K, then

$$\lim_{t \to \infty} (Kx)(t) = \int_0^\infty k(s) x(s) \, ds \tag{3.4}$$

for any $x \in C_g(R_+, R^n)$.

Proof   Let us show first that (3.2) and (3.3) are *sufficient* conditions for the admissibility of the pair (C_g, C_ℓ). We want to prove (3.4) for any $x \in C_g(R_+, R^n)$. From (3.2) it follows that the integral on the right-hand side of (3.4) makes sense for any $x \in C_g(R_+, R^n)$. Since

$$\lim_{t \to \infty} \int_0^t k(s) x(s) \, ds = \int_0^\infty k(s) x(s) \, ds \tag{3.5}$$

for any $x \in C_g(R_+, R^n)$, (3.4) can be written as

$$\lim_{t \to \infty} \int_0^t [k(t, s) - k(s)] x(s) \, ds = 0. \tag{3.6}$$

Taking into account that $\|x(t)\| \leq A_x g(t)$, $t \in R_+$, for some positive constant A_x, (3.6) will follow from

$$\lim_{t \to \infty} \int_0^t \|k(t, s) - k(s)\| g(s)\, ds = 0. \tag{3.7}$$

We shall now use (3.3) and hypothesis A to prove (3.7). For $0 < \tau \leq t$, we can write

$$\int_0^t \|k(t, s) - k(s)\| g(s)\, ds \leq \int_0^\tau \|k(t, s) - k(s)\| g(s)\, ds$$

$$+ \int_\tau^t \|k(t, s)\| g(s)\, ds + \int_\tau^t \|k(s)\| g(s)\, ds. \tag{3.8}$$

It is obvious that

$$\int_\tau^t \|k(s)\| g(s)\, ds \leq \int_\tau^\infty \|k(s)\| g(s)\, ds < \varepsilon \tag{3.9}$$

as long as $\tau \geq T_1(\varepsilon) > 0$. From (3.3) we derive

$$\int_0^t \|k(t, s)\| g(s)\, ds < \int_0^\infty \|k(s)\| g(s)\, ds + \varepsilon \tag{3.10}$$

for $t \geq T_2(\varepsilon) > 0$. Hypothesis A allows us to write

$$-\int_0^\tau \|k(t, s)\| g(s)\, ds < -\int_0^\tau \|k(s)\| g(s)\, ds + \varepsilon \tag{3.11}$$

as long as $t \geq T(\varepsilon, \tau) > 0$. From (3.9), (3.10), and (3.11), with $\tau = \max(T_1, T_2)$, we obtain

$$\int_\tau^t \|k(t, s)\| g(s)\, ds = \int_0^t \|k(t, s)\| g(s)\, ds - \int_0^\tau \|k(t, s)\| g(s)\, ds$$

$$< \int_0^\infty \|k(s)\| g(s)\, ds + \varepsilon - \int_0^\tau \|k(s)\| g(s)\, ds + \varepsilon$$

$$= \int_\tau^\infty \|k(s)\| g(s)\, ds + 2\varepsilon < 3\varepsilon \tag{3.12}$$

for $t \geq T(\varepsilon, \tau(\varepsilon)) = T_3(\varepsilon) > 0$. Finally, from hypothesis A we obtain

$$\int_0^\tau \|k(t, s) - k(s)\| g(s)\, ds < \varepsilon, \tag{3.13}$$

with τ fixed above and $t \geq T_4(\varepsilon) > 0$. Summing up our considerations we get

$$\int_0^t \|k(t, s) - k(s)\| g(s)\, ds < 5\varepsilon, \tag{3.14}$$

as long as $t \geq T(\varepsilon) = \max(T_1, T_2, T_3, T_4)$. From (3.14) we see that (3.7) holds. Consequently, the sufficiency of (3.2) and (3.3) is established.

To prove the *necessity* of these conditions we shall remark first that $Kx \in C_\ell(R_+, R^m)$ implies $Kx \in C(R_+, R^m)$. From the corollary to Theorem 2.1 one obtains

$$\int_0^t \|k(t,s)\| \|g(s)\| \, ds \leq M, \qquad t \in R_+, \tag{3.15}$$

where M is a positive constant. Since (3.15) can also be written in the form

$$\int_0^\infty \|k(t,s)\| \|g(s)\| \, ds \leq M, \qquad t \in R_+, \tag{3.16}$$

if we agree to extend $k(t, s)$ to the whole first quadrant such that $k(t, s) = 0$ for $s > t$, from Fatou's lemma applied to (3.16) we find

$$\int_0^\infty \|k(s)\| \|g(s)\| \, ds \leq M. \tag{3.17}$$

Therefore, condition (3.2) is necessary.

In order to prove the necessity of (3.3) we shall first restrict our considerations to the scalar case: $m = n = 1$. It is obvious that

$$L = \int_0^\infty |k(s)| \, |g(s)| \, ds \leq \liminf_{t \to \infty} \int_0^t |k(t,s)| \, |g(s)| \, ds$$

$$\leq \limsup_{t \to \infty} \int_0^t |k(t,s)| \, |g(s)| \, ds = L' \leq M. \tag{3.18}$$

It suffices to show that from $L' > L$ there results the existence of a function $x \in C_g(R_+, R)$ such that $\int_0^t k(t,s) x(s) \, ds \notin C_\ell(R_+, R)$. There exists a $t_0 \in R_+$ such that

$$\int_{t_0}^t |k(s)| \, |g(s)| \, ds < d < (L' - L)/3, \qquad t \geq t_0. \tag{3.19}$$

From

$$\int_{t_0}^t |k(t,s)| \, g(s) \, ds = \int_0^t |k(t,s)| \, g(s) \, ds - \int_0^{t_0} |k(t,s)| \, g(s) \, ds,$$

we obtain

$$\limsup_{t \to \infty} \int_{t_0}^t |k(t,s)| \, g(s) \, ds \geq L' - \int_0^{t_0} |k(s)| \, g(s) \, ds \geq L' - L > 3d.$$

Therefore, we can find an increasing sequence $\{t_n\}$, $t_n \to \infty$, such that

$$\int_{t_0}^{t_n} |k(t_n, s)| \, g(s) \, ds > 3d, \qquad n \geq 1. \tag{3.20}$$

From (3.19) there results

$$\int_{t_0}^{t_n} |k(s)| g(s)\, ds < d \tag{3.21}$$

for $n \geq 1$. We shall now prove that, without loss of generality, we can assume that the sequence $\{t_n\}$ also satisfies

$$\int_{t_0}^{t_n} |k(t_{n+1}, s)| g(s)\, ds < d, \qquad n \geq 1. \tag{3.22}$$

Indeed, from hypothesis A it follows that

$$\lim_{t \to \infty} \int_{t_0}^{t_1} |k(t, s)| g(s)\, ds = \int_{t_0}^{t_1} |k(s)| g(s)\, ds < d.$$

Consequently, we have

$$\int_{t_0}^{t_1} |k(t, s)| g(s)\, ds < d \tag{3.23}$$

for sufficiently large t. Hence, there exists a first $t_k > t_1$ in the sequence $\{t_n\}$ such that (3.23) holds for $t = t_k$. Let us omit all the terms of the sequence that lie between t_1 and t_k and denote t_k by t_2. We get

$$\int_{t_0}^{t_1} |k(t_2, s)| g(s)\, ds < d. \tag{3.24}$$

Starting now from

$$\lim_{t \to \infty} \int_{t_0}^{t_2} |k(t, s)| g(s)\, ds = \int_{t_0}^{t_2} |k(s)| g(s)\, ds < d,$$

we can construct t_3 such that

$$\int_{t_0}^{t_2} |k(t_3, s)| g(s)\, ds < d. \tag{3.25}$$

Of course, t_3 will also be chosen among the terms of the sequence $\{t_n\}$. Therefore, we can assume that the sequence $\{t_n\}$ satisfies both conditions (3.20) and (3.22).

The construction of a function $x \in C_g(R_+, R)$ such that $Kx \notin C_\ell(R_+, R)$ can be now easily accomplished. For each $n \geq 1$, let $x_n : [t_{n-1}, t_n) \to R$ be the mapping

$$x_n(t) = (-1)^{n-1} g(t)\, \text{sign}\, k(t_n, t). \tag{3.26}$$

It follows from (3.26) that x_n is measurable on $[t_{n-1}, t_n)$, that $|x_n(t)| \leq g(t)$ on the same interval, and that

$$(-1)^{n-1} \int_{t_{n-1}}^{t_n} k(t_n, s) x_n(s)\, ds > 2d. \tag{3.27}$$

Indeed,

$$(-1)^{n-1} \int_{t_{n-1}}^{t_n} k(t_n, s)x_n(s)\, ds = \int_{t_{n-1}}^{t_n} |k(t_n, s)|g(s)\, ds$$

$$= \int_{t_0}^{t_n} |k(t_n, s)|g(s)\, ds - \int_{t_0}^{t_{n-1}} |k(t_n, s)|g(s)\, ds$$

$$> 3d - d = 2d,$$

if we take into account (3.20) and (3.22). Using Lusin's theorem concerning the structure of measurable functions, we see that each $x_n(t)$ can be replaced by a continuous function $\tilde{x}_n(t)$ from $[t_{n-1}, t_n)$ into R such that the following conditions hold: $|\tilde{x}_n(t)| \leq g(t)$ on $[t_{n-1}, t_n)$, $\lim \tilde{x}_n(t) = \tilde{x}_{n+1}(t_n)$ as $t \to t_n$, and (3.27) with $\tilde{x}_n$ instead of x_n. Consider now a continuous mapping $x: R_+ \to R$ such that $x(t) = \tilde{x}_n(t)$ on $[t_{n-1}, t_n]$, $n \geq 1$, and $|x(t)| \leq g(t)$ on $[0, t_0)$. The existence of such a mapping is obvious if we take into account our previous considerations. We have $x \in C_g(R_+, R)$. It will be shown that

$$\lim_{t \to \infty} \int_{t_0}^{t} k(t, s)x(s)\, ds$$

does not exist, from which one derives that $Kx \notin C_\ell(R_+, R)$ because there exists

$$\lim_{t \to \infty} \int_0^{t_0} k(t, s)x(s)\, ds = \int_0^{t_0} k(s)x(s)\, ds$$

and

$$(Kx)(t) = \int_0^{t_0} k(t, s)x(s)\, ds + \int_{t_0}^{t} k(t, s)x(s)\, ds.$$

If we consider (3.7), we can write

$$(-1)^{n-1} \int_{t_0}^{t_n} k(t_n, s)x(s)\, ds = (-1)^{n-1} \int_{t_0}^{t_{n-1}} k(t_n, s)x(s)\, ds$$

$$+ (-1)^{n-1} \int_{t_{n-1}}^{t_n} k(t_n, s)x(s)\, ds$$

$$> 2d - \int_{t_0}^{t_{n-1}} |k(t_n, s)|x(s)\, ds$$

$$\geq 2d - \int_{t_0}^{t_{n-1}} |k(t_n, s)|g(s)\, ds > 2d - d = d.$$

This leads to the following inequalities:

$$\int_{t_0}^{t_n} k(t_n, s)x(s)\, ds > d \qquad \text{for} \quad n = 2p + 1,$$

$$\int_{t_0}^{t_n} k(t_n, s)x(s)\, ds < -d \qquad \text{for} \quad n = 2p.$$

Therefore, $\lim_{t \to \infty} \int_{t_0}^{t} k(t, s)x(s)\, ds$ cannot exist. This shows that the hypothesis $L' > L$ should be rejected and—consequently—(3.3) is necessary in the scalar case.

It remains to discuss the general case when $m, n \geq 1$. We notice that a vector function y belongs to $C_\ell(R_+, R^m)$ if and only if each coordinate of y belongs to $C_\ell(R_+, R)$. With the same procedure of reduction we used in the proof of Theorem 2.1 and taking into account the fact already established that (3.3) is necessary in the scalar case, we deduce that it suffices to prove the following statement. If

$$\lim_{t \to \infty} \int_0^t |k_{ij}(t, s)|\, g(s)\, ds = \int_0^\infty |k_{ij}(s)|\, g(s)\, ds$$

for $i = 1, 2, \ldots, m, j = 1, 2, \ldots, n$, then (3.3) holds. This follows immediately if we choose as norm for a matrix the sum of absolute values of its elements. Thus Theorem 3.1 is proved.

Remark Condition (3.2) can be replaced by condition (3.15). Indeed, from (3.15) and hypothesis A we derived (3.17), which is equivalent to (3.2). Conversely, from (3.2) and (3.3) we obtained the admissibility of the pair (C_g, C_ℓ) and—as shown above—this implies (3.15).

It is also obvious that (3.2) can be replaced by

$$\int_0^t \|k(t, s)\|\, g(s)\, ds \in C_\ell(R_+, R).$$

Corollary A necessary and sufficient condition that the pair (C_g, C_0) be admissible with respect to the operator K given by (2.1), with $k(t, s)$ satisfying hypothesis A, is that the following condition be fulfilled:

$$\int_0^t \|k(t, s)\|\, g(s)\, ds \in C_0(R_+, R). \qquad (3.28)$$

In equivalent form, (3.28) can be written as

$$\lim_{t \to \infty} \int_0^t \|k(t, s)\|\, g(s)\, ds = 0. \qquad (3.29)$$

The *sufficiency* of condition (3.28) follows easily from Theorem 3.1 (see also the remark to this theorem). Indeed, it implies both conditions (3.15) and (3.3), the latter with $k(s) \equiv 0$.

That (3.28) is *necessary* we can see as follows. From $x \in C_g$, there results $Kx \in C_0 \subset C_\ell$. Therefore, (3.3) and (3.4) are necessary. But we shall have $\lim_{t \to \infty}(Kx)(t) = 0$ for any $x \in C_g$, which implies that $\int_0^\infty k(s)x(s) \, ds = 0$. This leads easily to $k(s) \equiv 0$, and (3.3) reduces now to (3.29).

2.4 Admissibility of the Pair (L_g^∞, C_G)

First, we shall define the space $L_g^\infty = L_g^\infty(R, R^n)$, where $g: R \to R_+$ is measurable and positive. We set

$$L_g^\infty(R, R^n) = \{x : x/g \in L^\infty(R, R^n)\}, \tag{4.1}$$

the norm being given by

$$|x|_{L_g^\infty} = |x/g|_{L^\infty} = \text{ess sup}\{\|x(t)\|/g(t): t \in R\}. \tag{4.2}$$

The space L_g^∞ is obviously isometrically isomorphic to the space L^∞.

The operator we shall deal with in this section is

$$(Kx)(t) = \int_R k(t, s)x(s) \, ds, \qquad t \in R, \tag{4.3}$$

where $k(t, s)$ is a measurable matrix function on $R \times R$, of type m by n. It is our aim to find conditions for $k(t, s)$ such that the pair (L_g^∞, C_G) be admissible with respect to the operator K given by (4.3). Actually, we shall consider a special case of admissibility. Besides the admissibility condition $KL_g^\infty \subset C_G$, we shall assume that the following property holds:

B: For any bounded set $S \subset L_g^\infty$ and any $t_0 \in R$, the set $KS \subset C_G$ is equi-continuous at t_0. In other words, if $\varepsilon > 0$, there exists $\delta(\varepsilon) > 0$ such that

$$\|(Kx)(t) - (Kx)(t_0)\| < \varepsilon \qquad \text{when} \quad |t - t_0| < \delta$$

for any $x \in S \subset L_g^\infty$.

We can now prove a theorem giving necessary and sufficient conditions that the pair (L_g^∞, C_G) be admissible with respect to K and that property B hold true.

Theorem 4.1   Consider the spaces $L_g^\infty = L_g^\infty(R, R^n)$, $C_G = C_G(R, R^m)$ and the operator K given by (4.3), with $k(t, s)$ a measurable matrix function on $R \times R$ of type m by n. The necessary and sufficient conditions for admissibility of the pair $(L_g{}^\infty, C_G)$ with respect to the operator K, under the additional condition B, are:

$$\int_R \|k(t, s)\| g(s)\, ds \in C_G(R, R), \tag{4.4}$$

and

$$\lim_{t \to t_0} \int_R \|k(t, s) - k(t_0, s)\| g(s)\, ds = 0 \tag{4.5}$$

for any $t_0 \in R$.

Proof From (4.4) there results the existence of a positive constant M such that

$$\int_R \|k(t, s)\| g(s)\, ds \le MG(t), \qquad t \in R. \tag{4.6}$$

Assume that $x \in L_g{}^\infty(R, R^n)$. Then

$$\left\| \int_R k(t, s)x(s)\, ds \right\| \le \int_R \|k(t, s)\|\ \|x(s)\|\, ds$$

$$\le \int_R \|k(t, s)\| g(s)(\|x(s)\|/g(s))\, ds \le MG(t) |x|_{L_g{}^\infty},$$

which shows that $Kx \in C_G$ (the continuity easily follows from (4.5)). Therefore, conditions (4.4) and (4.5) are *sufficient*.

Let us now prove that they are also necessary. We shall again use the procedure of reduction that allows us to limit our considerations to the scalar case ($m = n = 1$). We fix $t \in R$ and consider the linear functional on $L_g{}^\infty$

$$L_t(x) = [G(t)]^{-1} \int_R k(t, s)x(s)\, ds. \tag{4.7}$$

Since $k(t, s)x(s) = k(t, s)g(s)y(s)$ with $y \in L^\infty$, and y is arbitrary in L^∞ when x is arbitrary in $L_g{}^\infty$, it follows that $L_t(x)$ is a continuous functional on $L_g{}^\infty$. Moreover, we have

$$\|L_t\| = [G(t)]^{-1} \int_R |k(t, s)| g(s)\, ds \tag{4.8}$$

for any $t \in R$. On the other hand, it follows that

$$\sup\{|L_t(x)| : x \in L_g{}^\infty\} < +\infty, \tag{4.9}$$

because $(Kx)(t) = G(t)L_t(x)$ and $Kx \in C_G$. Hence, the Banach–Steinhaus theorem applies to the family of linear operators $\{L_t(x) : t \in R\}$ from $L_g{}^\infty$ into R, which yields

$$\sup\{[G(t)]^{-1} \int_R |k(t, s)| g(s)\, ds : t \in R\} = M < +\infty. \tag{4.10}$$

Condition (4.10) is nothing but (4.4) for the case $m = n = 1$. Therefore, (4.4) is necessary. Assume now that (4.5) is not satisfied. Consequently, there exists at least one $t_0 \in R$ such that we can find a sequence $\{t_n\} \subset R$, with $\lim_{n \to \infty} t_n = t_0$, for which

$$\int_R |k(t_n, s) - k(t_0, s)| g(s) \, ds \geq \varepsilon_0 > 0, \qquad n \geq 1, \qquad (4.11)$$

where ε_0 is a fixed number. Now consider the sequence of functions $\{x_n\} \subset L_g^\infty(R, R)$, where

$$x_n(s) = g(s) \, \text{sign}[k(t_n, s) - k(t_0, s)], \qquad n \geq 1. \qquad (4.12)$$

It is obvious that the sequence $\{x_n\}$ is a bounded set in L_g^∞ (it belongs to the unit ball with center at the null element). We have

$$(Kx_n)(t_n) - (Kx_n)(t_0) = \int_R |k(t_n, s) - k(t_0, s)| g(s) \, ds, \qquad n \geq 1. \quad (4.13)$$

From (4.11) and (4.13) we see that the set $\{Kx_n\} \subset C_G$ cannot be equicontinuous at $t = t_0$. This contradicts property B, and thus the theorem is completely proved.

Remark 1 From condition (4.4), or its equivalent form (4.6), we actually obtained more than $KL_g^\infty \subset C_G$. The continuity of the operator K was established, as was the fact that its norm satisfies $\|K\| \leq M$, the same M appearing in (4.6).

Remark 2 We shall make some comments for the reader interested in linear systems theory. The triplet $(L_g^\infty, C_G; K)$ can be regarded as a model in the theory of linear systems. The elements of L_g^∞ are the *inputs*, and those of C_G are the *outputs*. The continuity of the operator K is the first requirement to be made for the stability of this system. In our paper [9], we called a linear system *strongly stable* if the operator K satisfies, besides continuity, property B. Hence, Theorem 4.1 gives necessary and sufficient conditions that the linear system $(L_g^\infty, C_G; K)$ be strongly stable.

It is worth pointing out that this condition of strong stability is automatically satisfied in a very general case of time-invariant systems. For instance, it holds for the system $(L^\infty, C; K)$ with K given by

$$(Kx)(t) = \int_R k(t - s)x(s) \, ds, \qquad t \in R,$$

where $k \in L(R, R)$. It is well known that the condition $k \in L(R, R)$ is the stability condition for the system $(L^\infty, C; K)$. Actually, it is also necessary and sufficient for the strong stability of the system.

Finally, we should like to mention the following feature of the system $(L_g^\infty, C_G; K)$: the fact that the set of outputs consists of continuous functions, while the inputs are only measurable, is consistent with the application needs. Indeed, it is always desirable to allow more freedom for the inputs and to get outputs with some smoothness properties.

2.5 Admissibility of the Pair (L^p, C_G), $1 < p < \infty$

A natural extension of the result from the preceding section is concerned with the admissibility of the pair (L^p, C_G) with respect to the operator

$$(Kx)(t) = \int_R k(t, s)x(s) \, ds, \qquad t \in R. \tag{5.1}$$

Of course, it would be possible to consider a space L_g^p instead of L^p, where L_g^p consists of all measurable functions for which $x/g^{1/p} \in L^p$, with g measurable and positive. Since no material changes appear when dealing with L_g^p spaces instead of L^p, we shall restrict our investigation to the admissibility problem of the pair (L^p, C_G) with respect to the operator K as given by (5.1). Besides admissibility, we will consider a property that can be formulated as follows.

B_p: For any bounded set $S \subset L^p$ and $t_0 \in R$, the set of functions $KS \subset C_G$ is equicontinuous at t_0.

Now we are able to formulate a new admissibility result. Let q be the conjugate exponent for p (i.e., $p^{-1} + q^{-1} = 1$).

Theorem 5.1   Let us consider the spaces $L^p = L^p(R, R^n)$, $C_G = C_G(R, R^m)$ and the operator K defined by (5.1), where $k(t, s)$ is a measurable matrix function on $R \times R$ of type m by n. Then (L^p, C_G) is admissible with respect to the operator K and property B_p holds, if and only if

$$\left\{ \int_R \|k(t, s)\|^q \, ds \right\}^{1/q} \in C_G(R, R) \tag{5.2}$$

and

$$\lim_{t \to t_0} \int_R \|k(t, s) - k(t_0, s)\|^q \, ds = 0 \tag{5.3}$$

for any $t_0 \in R$.

Proof   It is easy to see that (5.2) and (5.3) are sufficient in order to assure admissibility and property B_p. Hölder's inequality has to be applied.

The necessity of condition (5.2) can be obtained by means of the uniform boundedness principle, imitating the proof of the necessity of condition (4.4) from Theorem 4.1. Without loss of generality we can consider only the scalar case.

The family of functionals to be considered is formally given by (4.7). The norm of $L_t(x)$ will now be

$$\|L_t\| = [G(t)]^{-1} \left\{ \int_R |k(t, s)|^q \, ds \right\}^{1/q}, \tag{5.4}$$

and this leads to the scalar form of condition (5.2). A somewhat different argument is needed in order to prove the necessity of condition (5.3). If we denote by S the unit sphere in L^p centered at the origin, then, given an arbitrary $\varepsilon > 0$ and $t_0 \in R$, there corresponds $\delta = \delta(\varepsilon, t_0, S) > 0$ such that

$$|(Kx)(t) - (Kx)(t_0)| = \left| \int_R [k(t, s) - k(t_0, s)]x(s) \, ds \right| < \varepsilon \tag{5.5}$$

as long as $|t - t_0| < \delta$, for any $x \in S$. If we fix t such that $|t - t_0| < \delta$ is satisfied, then the mapping

$$x \to \int_R [k(t, s) - k(t_0, s)]x(s) \, ds \tag{5.6}$$

from L^p into R is continuous. Indeed, we obtain from (5.5)

$$\left| \int_R [k(t, s) - k(t_0, s)]x(s) \, ds \right| < \varepsilon |x|_{L^p} \tag{5.7}$$

for any $x \in L^p$. Therefore, the norm of the mapping (5.6) is at most ε, which means that

$$\left\{ \int_R |k(t, s) - k(t_0, s)|^q \, ds \right\}^{1/q} \le \varepsilon \tag{5.8}$$

whenever $|t - t_0| < \delta$. Hence (5.3) is necessary. Theorem 5.1 is thus proved.

Remark 1 Theorems 4.1 and 5.1 give necessary and sufficient conditions for the admissibility of the pair (L^p, C_G), $1 < p \le \infty$, with respect to the operator K, under the additional property B_p (respectively, B). The reader is invited to complete the discussion by considering the case $p = 1$.

Remark 2 The significance of condition B_p (respectively, B) is the following. First, it can be shown that for any bounded $S \subset L^p$, the set $KS \subset C_G$ is equicontinuous on any compact interval of R. Therefore, condition B_p (respectively, B) means that K is completely continuous from L^p into $C_c(R, R^m)$.

2.6   Admissibility of the Pair (L_g^∞, L_G^∞)

We shall discuss in this section the admissibility of the pair (L_g^∞, L_G^∞), with $L_g^\infty = L_g^\infty(R, R^n)$, $L_G^\infty = L_G^\infty(R, R^m)$, and the operator K given by

$$(Kx)(t) = \int_R k(t, s)x(s)\, ds, \qquad t \in R. \tag{6.1}$$

The kernel $k(t, s)$ is a measurable matrix function (on $R \times R$) of type m by n. It is assumed that both g and G are measurable on R and take only positive values.

Theorem 6.1   The pair of spaces (L_g^∞, L_G^∞) is admissible with respect to the integral operator (6.1) if and only if

$$\int_R \|k(t, s)\| g(s)\, ds \in L_G^\infty(R, R), \tag{6.2}$$

or, in equivalent form, if and only if there exists $M > 0$ such that

$$\int_R \|k(t, s)\| g(s)\, ds \le MG(t) \qquad \text{a.e. on } R. \tag{6.3}$$

Proof   We shall first prove that it suffices to discuss only the particular case $g(t) \equiv G(t) \equiv 1$. Indeed, we can write (6.1) in the form

$$[G(t)]^{-1}(Kx)(t) = \int_R [G(t)]^{-1} k(t, s)g(s)[x(s)/g(s)]\, ds.$$

If $x \in L_g^\infty$ implies $Kx \in L_G^\infty$, this means that

$$(\tilde{K}y)(t) = \int_R \tilde{k}(t, s)y(s)\, ds,$$

where $\tilde{k}(t, s) = [G(t)]^{-1} k(t, s)g(s)$ carries L^∞ into L^∞. The converse is also true. Of course, reduction is possible by changing the kernel $k(t, s)$ into $\tilde{k}(t, s)$.

   If we apply the usual reduction procedure from the vector case to the scalar one, we have to prove that a necessary and sufficient condition for the admissibility of the pair (L^∞, L^∞) with respect to the operator K given by (6.1) is

$$\int_R |k(t, s)|\, ds \le M \qquad \text{a.e. on } R, \tag{6.4}$$

with M a positive constant. Let us prove now the last statement.

   First, we shall remark that condition (6.4) is obviously a sufficient condition to assure $KL^\infty \subset L^\infty$. It implies even more, namely, that K is continuous from L^∞ into itself and $\|K\| \le M$.

Assume now that $x \in L^{\infty}(R, R)$ implies $Kx \in L^{\infty}(R, R)$. It is easy to see that this condition implies the continuity of the operator K from L^{∞} into itself. We shall apply the closed-graph theorem. Let $\{x_n\} \subset L^{\infty}$ be a sequence convergent to an element $x \in L^{\infty}$. Let $\{Kx_n\} \subset L^{\infty}$ also be convergent to a certain $y \in L^{\infty}$. We have to show that $y = Kx$, or more precisely, that $y(t) = (Kx)(t)$ almost everywhere on R. Indeed, we have

$$|(Kx_n)(t) - (Kx)(t)| \leq \int_R |k(t, s)| \, |x_n(s) - x(s)| \, ds \qquad (6.5)$$

for almost all $t \in R$. Since $x(t) \equiv 1 \in L^{\infty}$, we see that $\int_R k(t, s) \, ds \in L^{\infty}$. Therefore

$$\int_R |k(t, s)| \, ds < +\infty \qquad (6.6)$$

for almost all $t \in R$, say for $t \in R - E_0$, with mes $E_0 = 0$. For any $t \in R - E_0$, the integral appearing in the right member of (6.5) is finite. Moreover, it tends to zero as $n \to \infty$ because $|x_n(s) - x(s)| \to 0$ a.e. on R when $n \to \infty$, and Lebesgue's dominated convergence theorem can be applied (see (6.6)). Hence

$$\lim_{n \to \infty} (Kx_n)(t) = (Kx)(t) \qquad \text{a.e. on } R.$$

On the other hand, $Kx_n \to y$ in L^{∞} implies the existence of a subsequence that converges a.e. on R to y. Without loss of generality, we can assume that the sequence $\{Kx_n\}$ itself converges a.e. on R to y. The foregoing considerations lead to $y(t) = (Kx)(t)$ a.e. on R, which proves that the graph of the mapping $K: L^{\infty} \to L^{\infty}$ is closed. From the closed-graph theorem one obtains that $K: L^{\infty} \to L^{\infty}$ is continuous. From the continuity of K there results the existence of a positive constant M with the property

$$|Kx|_{L^{\infty}} \leq M|x|_{L^{\infty}}, \qquad x \in L^{\infty}. \qquad (6.7)$$

Condition (6.7) allows us to write

$$|(Kx)(t)| \leq M, \qquad t \in R - E_x, \qquad (6.8)$$

for any x such that $|x|_{L^{\infty}} \leq 1$, the set E_x being of measure zero. Let us now fix a $T > 0$ and consider the space $L^{\infty}([-T, T], R)$. It consists of the restrictions of all functions from $L^{\infty}(R, R)$ to the interval $[-T, T]$. For any $x \in L^{\infty}$ $([-T, T], R)$ we define $x = 0$ outside $[-T, T]$; then we get a function from $L^{\infty}(R, R)$ with the same norm. Hence, we can write

$$\left| \int_{|s| \leq T} k(t, s)x(s) \, ds \right| \leq M, \qquad t \in R - E_x, \qquad (6.9)$$

with mes $E_x = 0$, for any $x \in L^{\infty}([-T, T]R)$ such that $|x|_{L^{\infty}} \leq 1$. Let $\{x_n\} \subset L^{\infty}([-T, T], R)$ be a sequence such that $|x_n|_{L^{\infty}} \leq 1$, $n \geq 1$, and whose closure with respect to the convergence almost everywhere on $[-T, T]$ is the unit ball of $L^\infty([-T, T], R)$. Such a sequence can be constructed by considering the functions of the following structure: if r_k, $k = 1, 2, \ldots, p$, are rational numbers such that $-T < r_1 < r_2 < \cdots < r_p < T$, then x_n takes only rational values on each interval occurring in the above division of $[-T, T]$. We have

$$\left| \int_{|s| \leq T} k(t, s) x_n(s) \, ds \right| \leq M, \qquad t \in R - E, \tag{6.10}$$

with mes $E = 0$, E being the same for all $n \geq 1$. Indeed, from (6.9) we see that

$$\left| \int_{|s| \leq T} k(t, s) x_n(s) \, ds \right| \leq M, \qquad t \in R - E_{x_n},$$

with mes $E_{x_n} = 0$, $n \geq 1$. If we take $E = \bigcup_{n=1}^\infty E_{x_n}$, then (6.10) holds for any $n \geq 1$. Consider now a fixed t such that (6.6) and (6.10) hold. In other words, $t \in R - (E_0 \cup E)$. Let $x_t(s) = \operatorname{sign} k(t, s)$, $|s| \leq T$. There exists a subsequence $\{x_{n_p}\} \subset \{x_n\}$ such that $\lim_{p \to \infty} x_{n_p}(s) = x_t(s)$ a.e. on $[-T, T]$. One obtains for $t \in R - (E_0 \cup E)$

$$\int_{|s| \leq T} |k(t, s)| \, ds = \left| \int_{|s| \leq T} k(t, s) x_t(s) \, ds \right|$$

$$= \lim_{p \to \infty} \left| \int_{|s| \leq T} k(t, s) x_{n_p}(s) \, ds \right| \leq M, \tag{6.11}$$

if we take into account (6.6) and (6.10). One more step has to be done in order to get (6.4). Indeed, let $\{T_n\}$ be a sequence of positive numbers such that $T_n \to \infty$ as $n \to \infty$. To each T_n there corresponds a set E_n, with mes $E_n = 0$, such that for $t \in R - (E_0 \cup E_n)$ we have

$$\int_{|s| \leq T_n} |k(t, s)| \, ds \leq M. \tag{6.12}$$

(6.12) is nothing but (6.11) for $T = T_n$. From (6.12) we get

$$\int_R |k(t, s)| \, ds \leq M, \qquad t \in R - E', \tag{6.13}$$

where $E' = \bigcup_{n=1}^\infty E_n$ and mes $E' = 0$. The necessity of condition (6.4) is thus proved. This ends the proof of Theorem 6.1.

Remark   More sophisticated reasoning also gives

$$\|K\| = \operatorname{ess\,sup} \left\{ \int_R |k(t, s)| \, ds : t \in R \right\}. \tag{6.14}$$

It is useful to point out that $|(Kx)(t)| \leq \int_R |k(t, s)| \, ds$ a.e. on R, for any $x \in L^\infty(R, R)$ with $|x|_{L^\infty} \leq 1$. A result from Dunford and Schwartz [1] (see Chapter IV, Section 8, Theorem 23) provides the necessary tool in carrying out the proof of (6.14).

2.7 Admissibility with Respect to the Convolution Operator

The convolution operator is formally given by the formula

$$(Kx)(t) = \int_R k(s)x(t-s)\,ds, \qquad t \in R, \tag{7.1}$$

with $k(t)$ a measurable matrix function on R whose elements $k_{ij}(t)$ are real or complex-valued. A condition we shall assume valid throughout this section is

$$\int_R \|k(t)\|\,dt < +\infty. \tag{7.2}$$

The following theorem gives an idea of how "nice" the properties of the convolution operator are when (7.2) holds.

Theorem 7.1 Consider the convolution operator given by (7.1) and assume that $k(t)$ is an m by n matrix function satisfying (7.2). Then K is continuous from $E(R, R^n)$ to $E(R, R^m)$, where E stands for any one of the spaces M, S, P_ω, L^p with $1 \le p \le \infty$, C, AP, or A_ω. Moreover, it is completely continuous when $E = P_\omega$ or A_ω.

Proof Consider first the case $E = M$. If $x \in M$, then

$$\int_t^{t+1} \|(Kx)(u)\|\,du = \int_t^{t+1} \left\| \int_R k(s)x(u-s)\,ds \right\| du$$

$$\le \int_t^{t+1} \int_R \|k(s)\|\,\|x(u-s)\|\,ds\,du$$

$$= \int_R \|k(s)\|\,ds \int_t^{t+1} \|x(u-s)\|\,du \le |k|_L |x|_M,$$

taking into account condition (7.2). We put $|k|_L = \int_R \|k(s)\|\,ds$. Therefore $Kx \in M = M(R, R^m)$ and $\|Kx\|_M \le |k|_L |x|_M$. It follows also that $\|K\| \le |k|_L$. Let us remark that all the operations we performed are valid, since they follow from Tonelli's theorem (see Dunford and Schwartz [1, Chapter III, Section 11, Theorem 14]). The case $E = S$ presents no difficulty. We know that $S \subset M$ is a subspace of M, and it remains to show that $x \in S$ implies $Kx \in S$. This follows easily from the inequality

$$|(Kx)(t+\tau) - (Kx)(t)|_M \le |k|_L |x(t+\tau) - x(t)|_M, \tag{7.3}$$

which holds for any $x \in M$ and $\tau \in R$. For $\tau = \omega$ and $x \in P_\omega$ we get from (7.3) that $Kx \in P_\omega$. Therefore, the statement of the theorem holds true for $E = P_\omega$. The "complete continuity" part follows also from (7.3) and the compactness criterion in P_ω (see Section 1.1).

Let us now consider the case $E = L^p$. For $p = \infty$ the proof goes straight-forwardly and is left to the reader. For $p = 1$, Tonelli's theorem applies again: if $x \in L$, then

$$\int_R \|(Kx)(t)\| \, dt = \int_R \left\| \int_R k(s)x(t - s) \, ds \right\| dt$$

$$\leqslant \int_R \|k(s)\| \, ds \int_R \|x(t - s)\| \, dt = |k|_L |x|_L.$$

In other words, $|Kx|_L \leq |k|_L |x|_L$. Therefore, it remains to consider the case $1 < p < \infty$. Again using the standard reduction procedure to the scalar case, we can assume, without loss of generality, that all the functions involved in our considerations are scalar. Let $y \in L^q(R, R)$ be arbitrary, where q is the conjugate exponent with p: $p^{-1} + q^{-1} = 1$. Then

$$\left| \int_R (Kx)(t)y(t) \, dt \right| \leq \int_R \int_R |k(s)| \, |x(t - s)| \, |y(t)| \, ds \, dt$$

$$\leq \int_R |k(s)| \, ds \int_R |x(t - s)| \, |y(t)| \, dt$$

$$\leq \int_R |k(s)| \, ds \left(\int_R |x(t)|^p \, dt \right)^{1/p} \left(\int_R |y(t)|^q \, dt \right)^{1/q}.$$

Therefore, the linear functional $y \to \int_R (Kx)(t)y(t) \, dt$ is continuous on L^q. This implies

$$\left(\int_R |(Kx)(t)|^p \, dt \right)^{1/p} \leq \int_R |k(s)| \, ds \left(\int_R |x(t)|^p \, dt \right)^{1/p}$$

for any $x \in L^p$, which means that

$$|Kx|_{L^p} \leq |k|_L |x|_{L^p}. \tag{7.4}$$

The continuity of K is now established on any L^p, $1 \leq p \leq \infty$.

It is useful to point out that for $p = \infty$ we have

$$KL^\infty \subset C. \tag{7.5}$$

This follows from $KL^\infty \subset L^\infty$, if we also take into account that

$$\lim_{h \to 0} \int_R \|k(t + h) - h(t)\| \, dt = 0, \tag{7.6}$$

which is a consequence of $k \in L$.

From (7.5) we derive

$$KC \subset C, \tag{7.7}$$

the norm of K satisfying $\|K\| \le |k|_L$. It also follows that

$$K(AP) \subset AP, \tag{7.8}$$

observing that for any $x \in C$ we have

$$|(Kx)(t + \tau) - (Kx)(t)|_C \le |k|_L |x(t + \tau) - x(t)|_C. \tag{7.9}$$

Finally, the inclusion

$$KA_\omega \subset A_\omega \tag{7.10}$$

can be obtained without any difficulty. The property that K is completely continuous on A_ω follows easily from Ascoli's criterion and property (7.6) of the kernel. Theorem 7.1 is thereby proved.

Remark 1   No difficulty appears when considering vector-valued functions whose coordinates are complex.

Remark 2   From Theorem 7.1 it follows that K becomes completely continuous if we deal with its restriction to a very "thin" subspace (P_ω or A_ω). It would be interesting to know whether this property holds for richer spaces.

The following example shows that—generally—K is not completely continuous on AP. Hence, it cannot be completely continuous on C or on L^∞.

We shall consider the (scalar) kernel $k(t) = \exp\{-|t|\}$, $t \in R$, and the bounded sequence $\{\cos(t/n) : n \ge 1\}$. It follows that

$$\int_R e^{-|t-s|} \cos \frac{s}{n} \, ds = \frac{2n^2}{n^2 + 1} \cos \frac{t}{n}, \qquad n \ge 1. \tag{7.11}$$

The sequence appearing in the right-hand side of (7.11) has no subsequence that converges uniformly on R. Therefore, the operator K generated by the above kernel cannot be completely continuous on AP.

With the aim of establishing a result similar to Theorem 7.1 concerning the spaces C_ℓ and C_0, we shall consider the convolution operator on the positive half-axis:

$$(Kx)(t) = \int_0^\infty k(t - s)x(s) \, ds, \qquad t \in R_+, \tag{7.12}$$

where $k(t)$ satisfies condition (7.2) and $x \in C_\ell(R_+, R^n)$ or $x \in C_0(R_+, R^n)$.

Theorem 7.2   Let $k(t)$ be a kernel satisfying the assumptions of Theorem 7.1. Then the operator K given by (7.12) is continuous from $C_\ell(R_+, R^n)$ to $C_\ell(R_+, R^m)$. Moreover, $KC_0 \subset C_0$.

Proof From $C_\ell \subset C$ and Theorem 7.1, it follows that any Kx, with $x \in C_\ell$, is continuous and bounded on R_+. We shall now prove that

$$\lim_{t \to \infty} (Kx)(t) = \int_R k(s)x(\infty)\, ds \qquad (7.13)$$

for any $x \in C_\ell$, where $x(\infty) = \lim_{t \to \infty} x(t)$.

Let us remark that (7.12) can be also written in the form

$$(Kx)(t) = \int_{-\infty}^{t} k(s)x(t - s)\, ds, \qquad t \in R_+. \qquad (7.14)$$

On the other hand,

$$\int_R k(s)x(\infty)\, ds = \lim_{t \to \infty} \int_{-\infty}^{t} k(s)x(\infty)\, ds. \qquad (7.15)$$

Therefore, (7.13) will follow from

$$\lim_{t \to \infty} \int_{-\infty}^{t} k(s)[x(t - s) - x(\infty)]\, ds = 0. \qquad (7.16)$$

We can write for $t > T$

$$\left\| \int_{-\infty}^{t} k(s)[x(t - s) - x(\infty)]\, ds \right\| \leq \int_{-\infty}^{T} \| k(s) \|\, \| x(t - s) - x(\infty) \|\, ds$$

$$+ 2M \int_{T}^{t} \| k(s) \|\, ds, \qquad (7.17)$$

where $M \geq \| x(t) \|$ for $t \in R_+$. Let us fix T such that

$$\int_{T}^{t} \| k(s) \|\, ds < \varepsilon/4M \qquad \text{for} \quad t \geq T(\varepsilon), \qquad (7.18)$$

with $\varepsilon > 0$ given. Since

$$\int_{-\infty}^{T} \| k(s) \|\, \| x(t - s) - x(\infty) \|\, ds \leq \left(\int_R \| k(s) \|\, ds \right) \sup_{s \geq t - T} \| x(s) - x(\infty) \|,$$

it follows that

$$\int_{-\infty}^{T} \| k(s) \|\, \| x(t - s) - x(\infty) \|\, ds < \varepsilon/2, \qquad (7.19)$$

whenever

$$\sup_{s \geq t - T} \| x(s) - x(\infty) \| < \varepsilon/2 \left(\int_R \| k(s) \|\, ds \right). \qquad (7.20)$$

But (7.20) is true for sufficiently large t, say for $t \geq T_1$. From (7.17)–(7.19) we obtain

$$\left\| \int_{-\infty}^{t} k(s)[x(t-s) - x(\infty)]\, ds \right\| < \varepsilon, \tag{7.21}$$

wherever $t \geq \max(T, T_1)$. This means that (7.16) holds true, which yields (7.13). The statement of our theorem is proved for the space C_ℓ. Since $C_0 \subset C_\ell$, from (7.13) there results $KC_0 \subset C_0$.

Theorem 7.2 is thus proved, if we take into account that the continuity of K was already established in Theorem 7.1.

Remark If we consider the spaces $M = M(R_+, R^n)$, $L^p = L^p(R_+, R^n)$ with $1 \leq p \leq \infty$, and $C = C(R_+, R^n)$, then we get $KM(R_+, R^n) \subset M(R_+, R^n)$, etc. In other words, results similar to those encountered in Theorem 7.1 are valid for the spaces listed above in the case of functions defined on R_+.

2.8 Review of Other Admissibility Results

The mathematical literature provides a large number of admissibility results with respect to an integral operator. We are going to state here a few of these results and sketch the proofs of some.

Theorem 8.1 Assume that $k(t, s)$ is a measurable function from $R \times R$ into R such that

$$\operatorname*{ess\,sup}_{s \in R} \left\{ \int_R |k(t, s)|^p\, dt \right\}^{1/p} = M < +\infty, \tag{8.1}$$

where $p > 1$. Then the operator K defined by

$$(Kx)(t) = \int_R k(t, s) x(s)\, ds \tag{8.2}$$

is continuous from $L(R, R)$ to $L^p(R, R)$ and its norm is $\|K\| = M$.

Conversely, for any continuous operator K from $L(R, R)$ into $L^p(R, R)$, there exists a measurable kernel $k(t, s)$ satisfying (8.1) and such that K is given by (8.2).

Let q be the conjugate exponent to p: $p^{-1} + q^{-1} = 1$. We can write

$$\int_R |k(t, s)|\, |x(s)|\, ds = \int_R (|k|\, |x|^{1/p}) |x|^{1/q}\, ds$$

$$\leq \left\{ \int_R |k(t, s)|^p |x(s)|\, ds \right\}^{1/p} \left\{ \int_R |x(s)|\, ds \right\}^{1/q}.$$

Therefore,

$$\int_R \left(\int_R |k(t, s)| \, |x(s)| \, ds \right)^p dt \le (|x|_L)^{p/q} \int_R \int_R |k(t, s)|^p |x(s)| \, ds \, dt$$

$$\le (|x|_L)^{p/q} \int_R |x(s)| \, ds \int_R |k(t, s)|^p \, ds$$

$$\le M^p (|x|_L)^p,$$

since $(p/q) + 1 = p$. In other words,

$$|Kx|_{L^p} \le M|x|_L, \qquad x \in L, \tag{8.3}$$

which proves the continuity of K under condition (8.1).

The sufficiency of condition (8.1) is thereby proved. We shall not discuss the necessity part of Theorem 8.1.

Theorem 8.2   Consider the operator K given formally by (8.2) and assume that $k(t, s)$ is measurable, complex-valued, and such that

$$\operatorname*{ess\,sup}_{t \in R} \int_R |k(t, s)| \, ds \le M < +\infty \tag{8.4}$$

and

$$\operatorname*{ess\,sup}_{s \in R} \int_R |k(t, s)| \, dt \le M. \tag{8.5}$$

Then K is continuous from $L^p(R, \mathscr{C})$ into itself, for $1 \le p \le \infty$ and $\|K\| \le M$.

It follows from Theorem 6.1 that (8.4) guarantees the continuity of K from L^∞ into itself. From (8.5) we get easily that K is continuous from L into itself. Indeed, for $x \in L$ we have

$$\int_R dt \int_R |k(t, s)| \, |x(s)| \, ds \le \int_R |x(s)| \, ds \int_R |k(t, s)| \, dt \le M|x|_L,$$

which means that $|Kx|_L \le M|x|_L$.

At this point, a result of Riesz applies (see Dunford and Schwartz [1, Chapter VI, Section 10, Corollary 12]) and we obtain that K is continuous from $L^p(R, \mathscr{C})$ into itself for any p, $1 \le p \le \infty$.

It is interesting to point out that Theorem 8.2 generalizes a result we established in Section 2.7 for $k(t, s) = k(t - s)$, with $k \in L$.

The result we shall give now assures the complete continuity (compactness) of the operator K from L^p into itself.

Theorem 8.3 Let $1 < p < \infty$ and $q = p/(p-1)$. Assume that $k(t, s)$ is a measurable function from $R \times R$ into R such that

$$\left\{ \int_R \left(\int_R |k(t, s)|^p \, dt \right)^{q/p} ds \right\}^{1/q} = M < +\infty. \tag{8.6}$$

Then the operator K given by (8.2) is completely continuous from $L^p(R, R)$ into itself.

Let us remark that for $p = 2$, condition (8.6) becomes

$$\int_R \int_R |k(t, s)|^2 \, dt \, ds = M^2 < +\infty, \tag{8.7}$$

which is known as Hilbert–Schmidt condition. According to Theorem 8.3, the operator K defined by (8.2), with $k(t, s)$ satisfying (8.7), is completely continuous from $L^2(R, R)$ into itself.

We shall prove the last statement. That K is continuous if (8.7) is verified, one can easily check. Let us prove now that K carries the unit ball $\{x : |x|_{L^2} \leq 1\}$ into a relatively compact set. We have, uniformly on the unit ball,

$$\lim_{h \to 0} \int_R |(Kx)(t + h) - (Kx)(t)|^2 \, dt$$

$$\leq \lim_{h \to 0} \int_R \int_R |k(t + h, s) - k(t, s)|^2 \, dt \, ds = 0,$$

according to (8.7). Of course, we took into account that $|x|_{L^2} \leq 1$. We have further, uniformly in x, with $|x|_{L^2} \leq 1$,

$$\lim_{A \to \infty} \int_{|t| \geq A} |(Kx)(t)|^2 \, dt \leq \lim_{A \to \infty} \int_{|t| \geq A} \left\{ \int_R |k(t, s)|^2 \, ds \right\} dt = 0,$$

because $\int_R |k(t, s)|^2 \, ds$ is integrable on R, as shown by (8.7). The foregoing considerations lead to the conclusion that K carries the unit ball into a relatively compact set (see, for instance, Yosida [1, p. 275]).

A similar argument can be used in order to prove Theorem 8.3. We leave to the reader the task of carrying out the proof.

Another result we want to state here can be formulated as follows.

Theorem 8.4 Consider the integral operator K given by (8.2), and assume that $k(t, s)$ is a measurable matrix function (on $R \times R$) of type m by n such that

$$\int_R \|k(t, s)\| g(s) \, ds \in L^p(R, R), \tag{8.8}$$

where g is a positive measurable function. Then K is continuous from $L_g^\infty = L_g^\infty(R, R^n)$ to $L^p = L^p(R, R^m)$, $1 \le p \le \infty$.

The proof presents no difficulty and is left to the reader. Let us notice that for $p = \infty$, condition (8.8) becomes

$$\int_R \|k(t, s)\| g(s) \, ds \le M < +\infty \qquad \text{a.e. on } R.$$

We already encountered this condition in a more general form in Section 2.6.

2.9 Admissibility for Differential Systems

Consider the linear differential system

$$\dot{x} = A(t)x + u(t), \qquad t \in R_+, \tag{9.1}$$

where x is the unknown vector function (a mapping from R_+ into R^n), $A(t)$ is a square matrix of order n whose elements belong to $L_{loc}(R_+, R)$, and $u(t)$ is a vector function belonging to a certain function space (usually, a subspace of $L_{loc}(R_+, R^n)$).

The concept of admissibility with respect to the system (9.1) was introduced by Massera and Schäffer and was extensively studied in their book [1]. If $B = B(R_+, R^n)$ and $D = D(R_+, R^n)$ are Banach spaces of functions such that $B, D \subset L_{loc}(R_+, R^n)$ then the pair (B, D) is called *admissible* with respect to the system (9.1) when (and only when) $u \in B$ implies the existence of *at least one solution* $x \in D$ for (9.1). Let us notice that the concept of solution is meant here in the Carathéodory sense, i.e., $x(t)$ is absolutely continuous on any compact interval of R_+ and satisfies (9.1) almost everywhere on R_+. Therefore, $x(t)$ is always a continuous mapping from R_+ into R^n. Nevertheless, it is convenient to keep the space $L_{loc}(R_+, R^n)$ as underlying space.

In order to establish a relationship between the concept of admissibility with respect to an integral operator and that with respect to a differential system, we shall restrict our considerations to a particular case of admissibility for differential systems. Namely, the pair (B, D) will be called *fully admissible* for (9.1) if (and only if) all the solutions belong to D, for any $u \in B$.

Let $X(t)$ be the fundamental matrix of the homogeneous system $\dot{x} = A(t)x$, such that $X(0) = I$. Then

$$x(t) = X(t)x^0 + \int_0^t X(t)X^{-1}(s)u(s) \, ds, \qquad t \in R_+, \tag{9.2}$$

with $x^0 \in R^n$, gives all the solutions of (9.1).

We claim that the pair (B, D) is fully admissible for (9.1) if and only if the following conditions hold:

$$X(t)x^0 \in D \qquad \text{for any} \quad x^0 \in R^n, \tag{9.3}$$

and

$$\int_0^t X(t)X^{-1}(s)u(s)\,ds \in D \qquad \text{for any} \quad u \in B. \tag{9.4}$$

Indeed, if (B, D) is fully admissible, then (9.3) follows if we choose $u = \theta \in B$. (9.4) expresses that a certain solution (corresponding to $x(0) = 0 \in R^n$) of (9.1) belongs to D, no matter how we choose u in B. Consequently, both (9.3) and (9.4) are necessary. On the other hand, if they hold, from (9.2) we obtain easily that (B, D) is fully admissible for (9.1).

Let us consider now condition (9.4). It can obviously be read as the admissibility condition of the pair (B, D) with respect to the integral operator

$$(Ku)(t) = \int_0^t X(t)X^{-1}(s)u(s)\,ds, \qquad t \in R_+ . \tag{9.5}$$

Assume now that the Banach spaces B and D are stronger than $L_{\text{loc}}(R_+, R^n)$. Since $k(t, s) = X(t)X^{-1}(s)$ is continuous for $0 \le s \le t < +\infty$, one obtains without difficulty the continuity of K from $L_{\text{loc}}(R_+, R^n)$ into itself. Applying Lemma 1.1, it follows that K is continuous from B to D.

Summing up the preceding considerations, we get the following result:

Theorem 9.1 Consider the differential system (9.1), with $A(t)$ a square matrix of order n, whose elements belong to $L_{\text{loc}}(R_+, R)$. Let B, $D \subset L_{\text{loc}}(R_+, R^n)$ be some Banach spaces with topologies stronger than the topology of $L_{\text{loc}}(R_+, R^n)$. Then (B, D) is fully admissible for (9.1) if and only if any solution of the homogeneous system $\dot{x} = A(t)x$ belongs to D and the pair (B, D) is admissible with respect to the integral operator K given by (9.5). Moreover, if this property holds, K is continuous from B to D.

Several interesting applications of Theorem 9.1 can be made if we particularize the spaces B and D. For instance, taking $D = C$ and $B = L^p$, from Theorems 4.1 and 5.1 we obtain necessary and sufficient conditions that all the solutions of the system (9.1) be bounded on R_+ for any $u \in L^p$ (with $1 < p \le \infty$ and R_+ instead of R). Then there results

$$\|X(t)\| \le M < +\infty, \qquad t \in R_+ , \tag{9.6}$$

and

$$\int_0^t \|X(t)X^{-s}(s)\|^q \, ds \le N < +\infty, \qquad t \in R_+ , \tag{9.7}$$

where $q = p(p - 1)$ for $1 < p < \infty$ and $q = 1$ for $p = \infty$. In other words, conditions (9.6) and (9.7) are necessary and sufficient for the full admissibility of the pair (L^p, C) with respect to the system (9.1). It is obvious that condition (9.7) corresponds to condition (5.2), while condition (5.3) is automatically satisfied by the kernel $k(t, s) = X(t)X^{-1}(s)$ appearing in (9.5).

A noteworthy particular case of the above conditions is that corresponding to $q = 1$. The problem of finding conditions under which the pair (L^∞, C) is fully admissible with respect to the system (9.1) was first investigated by Perron (see Halanay [4] for a detailed discussion).

Let us point out that under additional assumptions concerning $A(t)$, sharper results can be obtained. For example, when $A(t)$ is bounded on R_+, condition (9.7) implies $\|X(t)X^{-1}(s)\| \le M e^{-\alpha(t-s)}$, $t \in R_+$, for some positive numbers M and α. This condition for $X(t)$ shows that the null solution of the system $\dot{x} = A(t)x$ is exponentially asymptotically stable. Therefore, the fact that the pair (L^p, C) is fully admissible has strong implications for the homogeneous system associated to (9.1). Further results of this kind can be found in Massera and Schäffer [1] and Coppel [1].

In concluding this section, we remark that other admissibility conditions established in the preceding sections could be applied in order to get results similar to Theorem 9.1. This task is left to the reader.

2.10   Integral Equations in C_G

Let us consider the integral equation

$$x(t) = h(t) + \int_0^t k(t, s)f(s; x) \, ds, \qquad t \in R_+, \qquad (10.1)$$

where $x = x(t)$ is the unknown function and $f(\,\cdot\,; x)$ is—generally—a nonlinear operator. More precisely $f(t; x)$ stands for $(fx)(t)$, the mapping $x \to fx$ being such that it carries any function from $C_G(R_+, R^m)$ into $C_g(R_+, R^n)$.

Since we are interested in getting conditions that assure the existence of a solution of Eq. (10.1) in the space $C_G(R_+, R^m)$, it is useful to point out the following feature of our equation. If the integral operator K, given by

$$(Kf)(t) = \int_0^t k(t, s)f(s) \, ds, \qquad t \in R_+, \qquad (10.2)$$

is such that $KC_g \subset C_G$, i.e., the pair (C_g, C_G) is admissible with respect to this operator, then the right member of (10.1) belongs to C_G for any $x \in C_G$ and given $h \in C_G$. Consequently, we can consider on $C_G(R_+, R^n)$ the operator

$$(Tx)(t) = h(t) + \int_0^t k(t, s)f(s; x) \, ds, \qquad (10.3)$$

which enjoys the property that any fixed point is a solution of (10.1). Of course, $k(t, s)$ has to be a matrix kernel of type m by n. Therefore, the existence problem for (10.1) is now reduced to the existence of fixed points for the operator T given by (10.3).

The following result is an easy consequence of the contraction mapping principle (see Exercise 2, Section 1.2) and of Theorem 2.1.

Theorem 10.1 Consider the integral equation (10.1) and assume that the following conditions hold:

1. $h(t) \in C_G(R_+, R^m)$;
2. $k(t, s)$ is a continuous matrix kernel of type m by n such that

$$\int_0^t \|k(t, s)\| g(s)\, ds \in C_G(R_+, R); \qquad (10.4)$$

3. the mapping $x \to fx$ from $C_G(R_+, R^m)$ into $C_g(R_+, R^n)$ is such that

$$|fx - fy|_{C_g} \leq \lambda |x - y|_{C_G}. \qquad (10.5)$$

Then there exists a unique solution in C_G of Eq. (10.1) whenever λ is sufficiently small.

Proof In the space $C_G(R_+, R^m)$, we will consider the operator T given by (10.3) and prove that it is a contraction operator for sufficiently small λ.

From (10.4) there results the admissibility of the pair (C_g, C_G) with respect to the operator K defined by (10.2). We can write (10.4) in the equivalent form

$$\int_0^t \|k(t, s)\| g(s)\, ds \leq AG(t), \qquad t \in R_+, \qquad (10.6)$$

where $A > 0$ is a constant. Therefore, if $x, y \in C_G(R_+, R^m)$, we obtain from (10.3):

$$(Tx)(t) - (Ty)(t) = \int_0^t k(t, s)[f(s; x) - f(s; y)]\, ds. \qquad (10.7)$$

Taking into account (10.6) and (10.7) we easily get

$$|Tx - Ty|_{C_G} \leq A |fx - fy|_{C_g} \qquad (10.8)$$

for any $x, y \in C_G$. If we now consider (10.5) and (10.8), there results

$$|Tx - Ty|_{C_G} \leq A\lambda |x - y|_{C_G} \qquad (10.9)$$

for any $x, y \in C_G$. If we assume

$$\lambda < A^{-1}, \qquad (10.10)$$

one obtains from (10.9) that T is a contraction operator on $C_G(R_+, R^m)$ and Theorem 10.1 is thereby proven.

Remark   If instead of condition (3) of Theorem 10.1 we assume that $x \to fx$ is a mapping from $\Sigma = \{x : x \in C_G(R_+, R^m), |x|_{C_G} \leq \rho\}$ into $C_G(R_+, R^m)$ such that (10.5) is satisfied for any $x, y \in \Sigma$, then we obtain an existence result in Σ provided $T\Sigma \subset \Sigma$ holds true. It can be easily seen that the inclusion $T\Sigma \subset \Sigma$ follows from

$$|h|_{C_G} + A|f\theta|_{C_G} \leq \rho(1 - \lambda A), \qquad (10.11)$$

where θ denotes the null element in $C_G(R_+, R^m)$.

Several corollaries will now be derived from Theorem 10.1 by conveniently choosing the various data involved in its statement.

Corollary 1   Consider Eq. (10.1) under the following conditions:

1. $h(t) \in C(R_+, R^m)$, i.e., $h(t)$ is a continuous bounded mapping from R_+ into R^m;

2. $k(t, s)$ is a continuous matrix kernel of type m by n such that

$$\int_0^t \|k(t, s)\| \, ds \leq A, \qquad t \in R_+; \qquad (10.12)$$

3. the mapping $(t, x) \to f(t, x)$ from $R_+ \times R^m$ into R^n is continuous and satisfies the Lipschitz condition:

$$\|f(t, x) - f(t, y)\| \leq \lambda \|x - y\|. \qquad (10.13)$$

Then there exists a unique continuous and bounded solution of Eq. (10.1).

The proof of Corollary 1 follows easily from that of Theorem 10.1 if we take into account the following circumstances: first, condition (10.12) is nothing but (10.4) with $g = G = 1$. Second, under condition (3) of Corollary 1 one obtains that $x \to fx$, where $(fx)(t) = f(t, x(t))$ is a mapping from $C(R_+, R^m)$ into $C(R_+, R^n)$.

Corollary 1 gives a boundedness result concerning Eq. (10.1). It can be easily applied to obtain a boundedness result for ordinary differential systems.

Corollary 2   Consider Eq. (10.1) and assume the following conditions:

1. $h(t)$ is a continuous mapping from R_+ into R^m, such that

$$\|h(t)\| \leq h_0 e^{-\beta t}, \qquad t \in R_+, \qquad (10.14)$$

where h_0 and β are positive numbers;

2. $k(t, s)$ is a continuous matrix kernel of type m by n such that

$$\|k(t, s)\| \leq K_0 e^{-\alpha(t-s)}, \qquad 0 \leq s \leq t < +\infty, \qquad (10.15)$$

with K_0 and α some positive constants and $\alpha > \beta$;

3. $f(t, x)$ is continuous for $t \in R_+$ and $x \in R^m$, takes its values in R^n, $f(t, 0) \equiv 0$, and satisfies (10.13).

Then there exists a unique solution of (10.1) such that

$$\|x(t)\| \le \rho\, e^{-\beta t}, \qquad t \in R_+, \tag{10.16}$$

for a certain $\rho > 0$, provided λ is small enough.

In order to derive Corollary 2 from Theorem 10.1, it is useful to remark that (10.14) means $h \in C_g(R_+, R^m)$, with $g(t) = e^{-\beta t}$. We obtain further that

$$\int_0^t \|k(t, s)\| e^{-\beta s}\, ds \le K_0 (\alpha - \beta)^{-1} e^{-\beta t}, \qquad t \in R_+, \tag{10.17}$$

if we consider (10.15). Therefore, the pair (C_g, C_g) with $g(t) = e^{-\beta t}$ is admissible with respect to the operator K generated by the kernel $k(t, s)$ satisfying (10.15). Finally, condition (3) of Corollary 2 leads immediately to the conclusion that $x \to fx$, $(fx)(t) = f(t, x(t))$, is an operator from $C_g(R_+, R^m)$ into $C_g(R_+, R^n)$ satisfying condition (10.5). Therefore, condition (3) from Theorem 10.1 is also satisfied, and this ends the proof of Corollary 2.

We shall now apply Corollary 2 to the ordinary differential system

$$\dot{x} = A(t)x + f(t, x), \qquad t \in R_+, \tag{10.18}$$

in order to derive the famous Poincaré–Liapunov theorem on asymptotic stability. It is well known (see, for instance, C. Corduneanu [12]) that (10.18) is equivalent to the integral equation

$$x(t) = X(t)x_0 + \int_0^t X(t)X^{-1}(s)f(s, x(s))\, ds, \tag{10.19}$$

where $X(t)$ is determined from $\dot{X} = A(t)X$, $X(0) = I$, and x_0 is an arbitrary vector. If we assume that the zero solution of the homogeneous system $\dot{x} = A(t)x$ is uniformly asymptotically stable, then there results

$$\|X(t)X^{-1}(s)\| \le K_0\, e^{-\alpha(t-s)}, \qquad 0 \le s \le t < +\infty,$$

which is nothing but (10.15). When $A(t)$ reduces to a constant matrix, the above situation occurs if and only if all the characteristic roots of A have negative real parts. For $f(t, x)$ we keep condition (3) of Corollary 2. It follows then that any solution of (10.18) satisfies an estimate of the form (10.16), i.e., the zero solution of (10.18) is exponentially asymptotically stable. A somewhat better result can be obtained if we take into account the remark we made for Theorem 10.1.

We shall again consider Eq. (10.1) and apply Theorem 1.2.1 in order to prove the existence of at least one solution in the space $C_G(R_+, R^m)$.

Let us now consider the ball

$$\Sigma = \{x : x \in C_G(R_+, R^m), \qquad |x|_{C_G} \le \rho\}, \qquad (10.20)$$

where $\rho > 0$ is given.

Theorem 10.2 Assume that conditions (1) and (2) from Theorem 10.1 are fulfilled. Assume further that the mapping $x \to fx$ is continuous from Σ, endowed with the topology induced by $C_c(R_+, R^m)$, into $C_g(R_+, R^n)$. If there exists $r > 0$ such that

$$|fx|_{C_g} \le r \qquad \text{for} \qquad x \in \Sigma, \qquad (10.21)$$

and if

$$|h|_{C_G} + Ar \le \rho \qquad (10.22)$$

holds true with A satisfying (10.6), then Eq. (10.1) has at least one solution belonging to Σ.

Proof We consider the space $C_c(R_+, R^m)$ as the underlying space and observe that Σ is a closed convex set in this space. Moreover, the operator T given by (10.3) is obviously defined on Σ and takes its values in $C_G(R_+, R^m)$. Indeed, for $x \in \Sigma$ we have $fx \in C_g(R_+, R^n)$, and, taking into account the fact that the pair (C_g, C_G) is admissible with respect to the integral operator K generated by the kernel $k(t, s)$, we obtain that the right member in (10.3) belongs to the space $C_G(R_+, R^m)$. Furthermore, inequality (10.22) implies $T\Sigma \subset \Sigma$. Since the mapping $x \to Tx$ is continuous from $C_c(R_+, R^m)$ into itself (actually, it is continuous from C_c into C_G), it remains to show that $T\Sigma$ is relatively compact in $C_c(R_+, R^m)$. In other words, we have to prove that the functions belonging to $T\Sigma$ are uniformly bounded and equicontinuous on any finite interval $[0, a]$, $a > 0$. We remark first that from $x \in \Sigma$ we derive $\|x(t)\| \le \rho G(t)$, $t \in R_+$. Therefore, the functions of Σ are uniformly bounded on any interval $[0, a]$, and so are those of $T\Sigma$ (because $T\Sigma \subset \Sigma$). Assume now that $0 \le t, u \le a$, given $a > 0$ and $x \in \Sigma$. Then

$$(Tx)(t) - (Tx)(u) = h(t) - h(u) + \int_0^t [k(t, s) - k(u, s)]f(s; x)\, ds$$

$$+ \int_u^t k(u, s) f(s; x)\, ds,$$

which leads easily to

$$\|(Tx)(t) - (Tx)(u)\| \le \|h(t) - h(u)\| + r \int_0^a \|k(t, s) - k(u, s)\| g(s)\, ds$$

$$+ r \int_u^t \|k(u, s)\| g(s)\, ds.$$

From the last inequality it follows that $T\Sigma$ is an equicontinuous set on $[0, a]$, if we consider the continuity properties of h, k, and g. Hence, Theorem 2.1 of Chapter 1 gives the desired result.

Remark 1   The condition concerning the continuity of the mapping $x \rightarrow fx$ can be obviously weakened. It suffices to assume that f is continuous with respect to the topology induced by C_c on both Σ and C_g. Then T is continuous from Σ into itself, the topology being that of uniform convergence on any compact interval of R_+ (see Lemma 2.1).

Remark 2   A similar result can be obtained if conditions (1) and (2) from Theorem 10.1 hold and $x \rightarrow fx$ is a completely continuous mapping from $C_G(R_+, R^m)$ into $C_g(R_+, R^n)$. Then T is a completely continuous operator from C_G into itself, and we can take C_G as underlying space. Generally, it seems easier to handle conditions related to the topology of C_c than conditions concerning the spaces C_G (first, we have in mind the condition of complete continuity).

2.11   Perturbed Integral Equations in C_G

It is easy to see that the linear Volterra equation

$$x(t) = h(t) + \int_0^t k(t, s)x(s) \, ds, \qquad t \in R_+, \tag{11.1}$$

has a unique solution $x \in C_c(R_+, R^n)$ for any h belonging to the same space. Of course, we assume that $k(t, s)$ is a continuous n by n matrix kernel. The method of successive approximations leads to the following formula for the solution of Eq. (11.1):

$$x(t) = h(t) + \int_0^t \gamma(t, s)h(s) \, ds, \qquad t \in R_+, \tag{11.2}$$

where $\gamma(t, s)$ is the resolvent kernel associated to $k(t, s)$. It is given by

$$\gamma(t, s) = \sum_{m=1}^{\infty} k_m(t, s), \qquad 0 \le s \le t < +\infty, \tag{11.3}$$

with

$$k_1(t, s) = k(t, s), \qquad k_{m+1}(t, s) = \int_s^t k_m(t, u)k(u, s) \, du, \qquad m \ge 1. \tag{11.4}$$

If we are interested in getting a solution belonging to $C_G(R_+, R^n)$ for any $h \in C_G(R_+, R^n)$, then formula (11.2) and Theorem 2.1 yield the following result: A necessary and sufficient condition that the solution of Eq. (11.1) belong to $C_G(R_+, R^n)$ for any h belonging to the same space is

$$\int_0^t \|\gamma(t, s)\| G(s)\, ds \in C_G(R_+, R). \tag{11.5}$$

Let us consider now the perturbed integral equation

$$x(t) = h(t; x) + \int_0^t k(t, s)x(s)\, ds, \qquad t \in R_+, \tag{11.6}$$

where—as usual—$h(t; x)$ stands for $(hx)(t)$. We will make suitable assumptions concerning the mapping $x \to hx$.

A noteworthy particularization for h is obtained for

$$h(t; x) = h(t) + \int_0^t k_0(t, s)x(s)\, ds. \tag{11.7}$$

From (11.6) and (11.7) we get

$$x(t) = h(t) + \int_0^t [k(t, s) + k_0(t, s)]x(s)\, ds, \tag{11.8}$$

which—compared with (11.1)—is an equation with a perturbed kernel.

The following existence result for (11.6) will be obtained by means of the contraction mapping principle.

Theorem 11.1   Consider Eq. (11.6) under the following conditions:

1. If $\gamma(t, s)$ denotes the resolvent kernel associated with $k(t, s)$, then (11.5) holds;

2. the mapping $x \to hx$ from $C_G(R_+, R^n)$ into itself satisfies the Lipschitz condition

$$|hx - hy|_{C_G} \le \lambda |x - y|_{C_G}. \tag{11.9}$$

Then there exists a unique solution in C_G for Eq. (11.6) whenever λ is sufficiently small.

Proof   We claim that Eq. (11.6) is equivalent to the functional-integral equation

$$x(t) = h(t; x) + \int_0^t \gamma(t, s)h(s; x)\, ds, \qquad t \in R_+. \tag{11.10}$$

Indeed, if Eq. (11.6) has a solution $x \in C_G(R_+, R^n)$, then it necessarily satisfies (11.10), because the solution of (11.1) is given by (11.2). Conversely, if x satisfies (11.10), then by direct calculation one obtains that it also satisfies

(11.6). The only fact to be considered is that $\gamma(t, s)$ verifies the so-called integral equation of the resolvent kernel

$$\gamma(t, s) = h(t, s) + \int_s^t k(t, u)\gamma(u, s)\, du, \tag{11.11}$$

which can be easily derived from (11.3) and (11.4).

Therefore, we can deal with Eq. (11.10) instead of (11.6). The operator

$$(Ux)(t) = h(t; x) + \int_0^t \gamma(t, s)h(s; x)\, ds \tag{11.12}$$

is defined on the whole space $C_G(R_+, R^n)$ and takes its values into the same space. If we show that there exists a unique fixed point for U, Theorem 11.1 will be proven.

Let us notice that condition (11.5) is equivalent to

$$\int_0^t \|\gamma(t, s)\| G(s)\, ds \le AG(t), \qquad t \in R_+, \tag{11.13}$$

for a suitable $A > 0$. If we take into account (11.9), from (11.12) we easily obtain

$$|Ux - Uy|_{C_G} \le \lambda(1 + A)|x - y|_{C_G} \tag{11.14}$$

for any $x, y \in C_G(R_+, R^n)$. Therefore, if

$$\lambda < (1 + A)^{-1},$$

Eq. (11.6) has a unique solution in C_G. Theorem 11.1 is thus proved.

Corollary 1 Consider the equation

$$x(t) = h(t) + \int_0^t \{k(t, s)x(s) + K(t, s, x(s))\}\, ds \tag{11.15}$$

and suppose that the following conditions hold:

1. $h \in C_G(R_+, R^n)$;
2. the resolvent kernel $\gamma(t, s)$ associated to $k(t, s)$ satisfies (11.5);
3. the function $K(t, s, x)$ is continuous for $0 \le s \le t < +\infty$, $x \in R^n$, takes its values in R^n, $K(t, s, 0) \equiv 0$, and

$$\|K(t, s, x) - K(t, s, y)\| \le k_0(t, s)\|x - y\|, \tag{11.16}$$

where $k_0(t, s)$ is a positive continuous function for $0 \le s \le t < +\infty$, satisfying

$$\int_0^t k_0(t, s)G(s)\, ds \le M_0\, G(t), \qquad t \in R_+. \tag{11.17}$$

Then, there exists a unique solution of Eq. (11.15) belonging to $C_G(R_+, R^n)$ whenever M_0 is small enough.

The proof of Corollary 1 follows immediately from Theorem 11.1 if we notice that the operator

$$h(t; x) = h(t) + \int_0^t K(t, s, x(s))\, ds$$

satisfies the conditions required by that theorem (in particular, M_0 will play the role of the Lipschitz constant).

Corollary 2   Consider Eq. (11.1) and (11.8) under the following conditions:

1. for any bounded (and continuous) $h(t)$, the solution of (11.1) is also bounded on R_+;
2. the perturbing (matrix) kernel $k_0(t, s)$ satisfies

$$\int_0^t \|k_0(t, s)\|\, ds \le M_0, \qquad t \in R_+. \tag{11.18}$$

Then, Eq. (11.8) has a unique bounded solution for any bounded $h(t)$, provided M_0 is sufficiently small.

Indeed, condition (1) states that

$$\int_{0.}^t \|\gamma(t, s)\|\, ds \le A, \qquad t \in R_+, \tag{11.19}$$

and condition (2) leads to the conclusion that $h(t; x)$ given by (11.7) acts from $C_G(R_+, R^n)$ into itself and satisfies a Lipschitz condition with constant M_0. Therefore, Theorem 11.1 applies and Corollary 2 is proven.

In other words, Corollary 2 states that the property of Eq. (11.1) of possessing a bounded solution for any bounded free term is preserved when perturbing the kernel by another kernel that satisfies (11.18), with small M_0.

2.12   Existence of Convergent Solutions

By *convergent solution* of an integral (differential) equation we shall mean a solution belonging to a space C_ℓ. Let us consider the integral equation

$$x(t) = h(t) + \int_0^t k(t, s)f(s; x)\, ds, \qquad t \in R_+, \tag{12.1}$$

where $x = x(t)$ is an unknown vector function and $f(t; x) = (fx)(t)$ stands for an operator. Besides continuity, we will assume throughout this section that $k(t, s)$ satisfies *hypothesis A* from the Section 2.3, i.e.,

$$\lim_{t \to \infty} k(t, s) = k(s), \qquad s \in R_+, \tag{12.2}$$

uniformly on any compact interval of R_+.

Theorem 12.1 Assume that Eq. (12.1) satisfies the following conditions:

1. $h(t) \in C_\ell(R_+, R^n)$;
2. the pair $(C_g(R_+, R^m), C_\ell(R_+, R^n))$ is admissible with respect to the operator

$$(Kf)(t) = \int_0^t k(t, s)f(s)\, ds, \qquad t \in R_+, \tag{12.3}$$

i.e., conditions (3.2) and (3.3) are satisfied (see Theorem 3.1);

3. the mapping $x \to fx$ from $C(R_+, R^n)$ into $C_g(R_+, R^m)$ is such that

$$|fx - fy|_{C_g} \le \lambda |x - y|_C \tag{12.4}$$

holds for any $x, y \in C_\ell$. Then there exists a unique solution $x \in C_\ell(R_+, R^n)$ of Eq. (12.1) whenever λ is sufficiently small.

Proof It follows easily from our hypotheses that the operator

$$(Tx)(t) = h(t) + \int_0^t k(t, s)f(s; x)\, ds \tag{12.5}$$

carries the space $C_\ell(R_+, R^n)$ into itself. Consequently, if we can show that T is a contraction (of course, in the metric of C), the proof of Theorem 12.1 will be accomplished.

We know from Theorem 3.1 that the operator K given by (12.3) is continuous from $C_g(R_+, R^m)$ into $C_\ell(R_+, R^n)$. Therefore, we can find a positive constant M such that $|Kf|_C \le M|f|_{C_g}$ for any $f \in C_g(R_+, R^m)$. This leads easily to

$$|Tx - Ty|_C \le \lambda M |x - y|_C \tag{12.6}$$

for any $x, y \in C_\ell$. For $\lambda < M^{-1}$ we get from (12.6) that T is a contraction, and this ends the proof.

Remark It can be proven that

$$x(\infty) = h(\infty) + \int_0^\infty k(s)f(s; x)\, ds, \tag{12.7}$$

where $x(\infty) = \lim_{t \to \infty} x(t)$. Indeed, both $x(\infty)$ and $h(\infty)$ exist. Taking into account the last statement of Theorem 3.1, we easily obtain (12.7).

We shall now apply the result of the preceding theorem to a differential system of the form

$$\dot{x} = A(t)x + f(t, x), \qquad t \in R_+, \tag{12.8}$$

where x is a mapping from R_+ into R^n, $A(t)$ is a square matrix function of type n by n whose elements are continuous on R_+, and $(t, x) \to f(t, x)$ is a function from $R_+ \times R^n$ into R^n.

Let $X(t)$ be the fundamental matrix of the system $\dot{x} = A(t)x$ such that $X(0) = I$.

Theorem 12.2 Consider the differential system (12.8) and assume that:

1. there exists $\lim_{t \to \infty} X(t) = X(\infty)$, i.e., any element of $X(t)$ belongs to $C_\ell(R_+, R)$;
2. we have

$$\int_0^\infty \|X^{-1}(s)\|g(s)\, ds < +\infty \tag{12.9}$$

for a suitable continuous and positive function $g(t)$;
3. the mapping $(t, x) \to f(t, x)$ is continuous, $f(t, 0) \in C_g(R_+, R^n)$, and

$$\|f(t, x) - f(t, y)\| \le \lambda g(t)\|x - y\| \tag{12.10}$$

for any $x, y \in R^n$.

Then any solution of the system (12.8) is convergent, provided λ is small enough.

Proof Since system (12.8) is equivalent to the family of integral equations

$$x(t) = X(t)x^0 + \int_0^t X(t)X^{-1}(s)f(s, x(s))\, ds \tag{12.11}$$

with $x^0 \in R^n$, we have to verify that Eq. (12.11) satisfies the hypotheses of Theorem 12.1. Indeed, condition (1) from our theorem implies the corresponding condition (1) from the statement of Theorem 12.1. From conditions (1) and (2) of our theorem we can easily derive the admissibility of the pair (C_g, C_ℓ) with respect to the integral operator K given by

$$(Kf)(t) = \int_0^t X(t)X^{-1}(s)f(s)\, ds. \tag{12.12}$$

More precisely, we have to prove that

$$\lim_{t \to \infty} \int_0^t \|X(t)X^{-1}(s)\|g(s)\, ds = \int_0^\infty \|X(\infty)X^{-1}(s)\|g(s)\, ds, \tag{12.13}$$

the integral in the right member of (12.13) being obviously convergent according to our condition (2). Some elementary calculations show that

$$\left| \int_0^t \|X(t)X^{-1}(s)\|g(s)\, ds - \int_0^\infty \|X(\infty)X^{-1}(s)\|g(s)\, ds \right|$$

$$\le \|X(t) - X(\infty)\| \int_0^t \|X^{-1}(s)\|g(s)\, ds + \|X(\infty)\| \int_t^\infty \|X^{-1}(s)\|g(s)\, ds,$$

and this leads easily to (12.13). Finally, the operator f defined by $(fx)(t) = f(t, x(t))$ acts from $C_\ell(R_+, R^n)$ into $C_g(R_+, R^n)$ and satisfies a Lipschitz condition with constant λ. The last statement follows easily from (12.10). Theorem 12.2 is thereby proved.

Remark 1 It is obvious that we can assume $A(t)$ locally integrable on R_+ (not necessarily continuous). Solutions of the system (12.8) should then be considered in the Carathéodory sense.

Remark 2 Conditions (1) and (2) of Theorem 12.2 are necessary and sufficient for the full admissibility of the pair (C_g, C_ℓ) with respect to the differential system $\dot{x} = A(t)x + u(t)$. This follows from the considerations expounded in Section 2.9.

We are going to consider Eq. (12.1) again under conditions that guarantee the existence of at least one convergent solution. The Schauder–Tychonoff fixed-point principle will be used. The following compactness criterion in $C_\ell(R_+, R^n)$ is particularly useful.

Let $\mathscr{F} \subset C_\ell(R_+, R^n)$ be a set satisfying the following conditions:

1. $\mathscr{F}$ is bounded in C_ℓ;
2. the functions belonging to $\mathscr{F}$ are equicontinuous on any compact interval of R_+;
3. the functions from $\mathscr{F}$ are equiconvergent, i.e., given $\varepsilon > 0$, there corresponds $T(\varepsilon) > 0$ such that $\|f(t) - f(\infty)\| < \varepsilon$ for any $t \geq T(\varepsilon)$ and $f \in \mathscr{F}$.

Then $\mathscr{F}$ is compact in $C_\ell(R_+, R^n)$.

We will sketch the proof of the above criterion. First, from any sequence of $\mathscr{F}$ we can extract a subsequence, say $\{f^m\}$, such that $\{f^m(\infty)\}$ is convergent. From condition (3) it follows that

$$\|f^m(t) - f^m(\infty)\| < \varepsilon/3 \qquad \text{if} \quad t \geq T(\varepsilon), \quad m \geq 1. \qquad (12.14)$$

If m, p are sufficiently large, then

$$\|f^m(\infty) - f^p(\infty)\| < \varepsilon/3. \qquad (12.15)$$

Assume that (12.15) holds for $m, p \geq N(\varepsilon)$. From (12.14) and (12.15) we derive

$$\|f^m(t) - f^p(t)\| < \varepsilon \qquad (12.16)$$

as long as $t \geq T(\varepsilon)$ and $m, p \geq N(\varepsilon)$. On the other hand, $\{f^m(t)\}$ is uniformly bounded and equicontinuous on the interval $[0, T]$, as stated in conditions (1) and (2). Therefore, $\{f^m(t)\}$ contains a subsequence that is uniformly convergent on $[0, T]$, say $\{f^{m_k}(t)\}$. We can write

$$\|f^{m_k}(t) - f^{m_h}(t)\| < \varepsilon \qquad (12.17)$$

for $t \in [0, T]$ and m_k, $m_h \geq N_1(\varepsilon)$. Now, from (12.16) and (12.17), we see that $\{f^{m_k}(t)\}$ satisfies the Cauchy's condition on R_+. Hence, it is uniformly convergent on R_+, i.e., it converges in C_ℓ.

The above criterion leads easily to the following result concerning the integral operators acting from C_g to C_ℓ.

Lemma 12.1   Consider the integral operator K defined by (12.3), with $k(t, s)$ continuous, satisfying hypothesis A, and such that the pair (C_g, C_ℓ) is admissible. Then K is completely continuous.

Proof   Let $M \subset C_g$ be a bounded set. This means that $\|f(t)\| \leq Ng(t)$, $t \in R_+$, for any $f \in M$, with fixed $N > 0$. Denote by $\mathscr{F}$ the set $\{Kf : f \in M\}$. We have to show that $\mathscr{F}$ satisfies the conditions listed in the above criterion of compactness. The boundedness of $\mathscr{F}$ is a consequence of the continuity of K. Condition (2) of the criterion follows easily if we take into account the inequality

$$\|(Kf)(t) - (Kf)(u)\| \leq N \int_0^t \|k(t, s) - k(u, s)\| g(s) \, ds + N \left| \int_u^t \|k(u, s)\| g(s) \, ds \right|.$$

Finally, condition (3) follows from

$$\|(Kf)(t) - (Kf)(\infty)\| \leq N \int_0^t \|k(t, s) - k(s)\| g(s) \, ds + N \int_t^\infty \|k(s)\| g(s) \, ds,$$

if we consider (3.2) and (3.7). This ends the proof of Lemma 12.1.

Theorem 12.3   Assume that the following conditions hold for Eq. (12.1):

1. and 2.—the same as in Theorem 12.1;
3. the mapping $x \to fx$ is continuous from

$$\Sigma = \{x : x \in C_\ell(R_+, R^n), \, |x|_C \leq \rho\}$$

into $C_g(R_+, R^m)$ and

$$|fx|_{C_g} \leq r, \qquad x \in \Sigma; \tag{12.18}$$

4. if M denotes the norm of K, then

$$|h|_C + Mr \leq \rho. \tag{12.19}$$

Then there exists at least one solution of Eq. (12.1) such that $x \in \Sigma$.

Proof   The operator T given by (12.5) is obviously continuous on Σ and carries this set into itself. Indeed,

$$|Tx|_C \leq |h|_C + \left| \int_0^t k(t, s) f(s; x) \, ds \right|_C \leq |h|_C + M |fx|_{C_g} \leq |h|_C + Mr \leq \rho,$$

which shows that $T\Sigma \subset \Sigma$. Since K is completely continuous from C_g to C_ℓ, it follows that T carries Σ into a relatively compact set in C_ℓ. On the other hand, Σ is a convex closed set in C_ℓ and this implies the existence of at least one fixed point for T in Σ. Therefore, Theorem 12.3 is proved.

We shall obtain further existence results in C_ℓ in Section 2.14, where convolution equations are investigated.

2.13   Further Existence Results in C_G

The integral equations we considered in the preceding section involve a linear integral operator of Volterra type. In this section we are going to study some integral equations of the form

$$x(t) = h(t) + \int_0^\infty k(t, s) f(s; x)\, ds, \qquad t \in R_+ , \tag{13.1}$$

under suitable conditions that assure the existence of a continuous solution on R_+. The results concerning the integral operators we established in Sections 2.4 and 2.5 are needed.

Theorem 13.1   Assume that the following conditions hold for Eq. (13.1):

1. $h(t) \in C_G(R_+ , R^n)$ with $G(t)$ a continuous positive function on R_+;
2. $k(t, s)$ is a measurable matrix kernel of type n by m such that

$$\int_0^\infty \|k(t, s)\| g(s)\, ds \in C_G(R_+ , R) \tag{13.2}$$

with $g(t)$ measurable and positive on R_+, and

$$\lim_{t \to t_0} \int_0^\infty \|k(t, s) - k(t_0 , s)\| g(s)\, ds = 0 \tag{13.3}$$

for any $t_0 \in R_+$;
3. the mapping $x \to fx$ is continuous from

$$\Sigma = \{x : x \in C_G(R_+ , R^n),\ |x|_{C_G} \le \rho\},$$

endowed with the topology induced by $C_c(R_+ , R^n)$, into $L_g(R_+ , R^m)$ and there exists $r > 0$ such that

$$|fx|_{L_g^\infty} \le r, \qquad x \in \Sigma. \tag{13.4}$$

Then there exists at least one solution $x \in \Sigma$ of Eq. (13.1), provided $|h|_{C_G}$ and r are sufficiently small.

Proof   We assume that $C_c(R_+, R^n)$ is the underlying space. The set Σ is then convex and closed. Consider the operator on Σ given by

$$(Tx)(t) = h(t) + \int_0^\infty k(t, s) f(s; x) \, ds, \qquad t \in R_+. \tag{13.5}$$

It suffices to show that T is continuous from $C_c(R_+, R^n)$ into itself, $T\Sigma \subset \Sigma$, and $T\Sigma$ is relatively compact in $C_c(R_+, R^n)$. First, $x \to fx$ is continuous from $C_c(R_+, R^n)$ into $L_g^\infty(R_+, R^m)$ according to condition (3) of Theorem 13.1. Second, $f \to Kf$ is continuous from $L_g^\infty(R_+, R^m)$ into $C_G(R_+, R^n)$. Hence $x \to Kfx$ is continuous from C_c into C_G and this implies, of course, that $x \to Kfx$ is continuous from C_c into itself. Let M be such that $|Kf|_{C_G} \leq M|f|_{L_g^\infty}$. Then $x \in \Sigma$ implies

$$|Tx|_{C_G} \leq |h|_{C_G} + Mr \leq \rho \tag{13.6}$$

as long as $|h|_{C_G}$ and r are sufficiently small. Therefore, $T\Sigma \subset \Sigma$ if (13.6) holds. The only point that still needs discussion is showing that $T\Sigma$ is relatively compact in C_c. In other words, we have to show that the functions belonging to $T\Sigma$ are uniformly bounded and equicontinuous on any compact interval of R_+. The uniform boundedness follows from $T\Sigma \subset \Sigma$, which implies $\|(Tx)(t)\| \leq \rho G(t)$, $t \in R_+$, for any $x \in \Sigma$. The equicontinuity property is a consequence of Theorem 4.1. This ends the proof of Theorem 13.1.

Remark 1   By means of the Banach contraction mapping principle, one can easily prove a result of existence and uniqueness in C_G for Eq. (13.1). The task of formulating the corresponding conditions is left to the reader.

Remark 2   If $(fx)(t) = f(t, x(t))$, then the classical Carathéodory conditions (i.e., measurability with respect to t and continuity in x) are required to ensure condition (3) of the theorem. Of course, the boundedness property (13.4) should also be preserved.

Another existence result concerning Eq. (13.1) will be obtained under the main hypothesis that the pair (L^p, C_G), $1 < p < \infty$, is admissible with respect to the linear integral operator K generated by the kernel $k(t, s)$.

Theorem 13.2   Consider Eq. (13.1) and assume that the following conditions hold:

1. $h(t) \in C_G(R_+, R^n)$;
2. the pair (L^p, C_G), $1 < p < \infty$, is admissible for K, and property B_p stated in Section 2.5 holds true;
3. the mapping $x \to fx$ from $C_G(R_+, R^n)$ into $L^p(R_+, R^m)$ satisfies the Lipschitz condition

$$|fx - fy|_{L^p} \leq \lambda |x - y|_{C_G}. \tag{13.7}$$

Then there exists a unique solution $x \in C_G(R_+, R^n)$ of Eq. (13.1), provided λ is small enough.

Proof The contraction mapping principle can be applied in the space $C_G(R_+, R^n)$. We notice that Theorem 5.1 also gives an estimate for the norm of operator K, a feature that allows us to determine how small λ should be in (13.7).

Remark By means of the Schauder–Tychonoff fixed-point theorem we can prove a result very similar to Theorem 13.1. This task is left to the reader.

2.14 Convolution Equations

In this section we shall use the results presented in Section 2.7 for the linear convolution operator. We should like to point out that, under various kinds of assumptions, the convolution equations will be the object of our investigation in Chapters 3 and 4. The main goal of the present section is to establish some existence results related to the linear convolution equation

$$x(t) = h(t) + \int_R k(s)x(t - s) \, ds, \qquad t \in R, \tag{14.1}$$

and also to the Hammerstein equation associated with (14.1):

$$x(t) = h(t) + \int_R k(s)f(t - s; x) \, ds, \qquad t \in R. \tag{14.2}$$

In order to obtain conditions under which Eq. (14.1) has a unique solution belonging to a certain function space E for any $h \in E$, we shall need the admissibility results from Section 2.7, and also Theorem 1.3.5 on Fourier transforms. We shall discuss only the scalar case.

Theorem 14.1 Assume that the following conditions hold for (14.1):

1. $k \in L(R, \mathscr{C})$;
2. if $\tilde{k}(s) = \int_R k(t)\exp\{its\} \, dt$, then

$$\tilde{k}(s) \neq 1, \qquad s \in R. \tag{14.3}$$

Then there exists a unique solution $x \in E$ of Eq. (14.1) for any $h \in E$, where E stands for any one of the spaces, $M, S, P_\omega, L^p(1 \leq p \leq \infty), C, AP,$ or A_ω.

Proof Assume first that $E = L(R, \mathscr{C})$. Then we can take the Fourier transform of both sides in (14.1) and obtain

$$\tilde{x}(s) = \tilde{h}(s) + \tilde{k}(s)\tilde{x}(s), \qquad s \in R. \tag{14.4}$$

If we take into account (14.3), there results

$$\tilde{x}(s) = [1 - \tilde{k}(s)]^{-1}\tilde{h}(s), \qquad s \in R. \qquad (14.5)$$

Now, from Theorem 1.3.5 we derive that

$$[1 - \tilde{k}(s)]^{-1} = 1 + \tilde{k}_1(s), \qquad s \in R, \qquad (14.6)$$

for a certain $k_1 \in L(R, \mathscr{C})$. Hence, (14.5) can be written in the form

$$\tilde{x}(t) = \tilde{h}(s) + \tilde{k}_1(s)\tilde{h}(s), \qquad s \in R. \qquad (14.7)$$

Therefore, there exists $k_1 \in L(R, \mathscr{C})$ such that

$$x(t) = h(t) + \int_R k_1(s)h(t - s)\, ds, \qquad t \in R, \qquad (14.8)$$

for any $h \in L(R, \mathscr{C})$.

Since (14.6) can be written as

$$\tilde{k}_1(s) = \tilde{k}(s) + \tilde{k}(s)\tilde{k}_1(s), \qquad s \in R, \qquad (14.9)$$

it follows that

$$k_1(t) = k(t) + \int_R k(t - s)k_1(s)\, ds, \qquad t \in R. \qquad (14.10)$$

Equation (14.10) is called the integral equation of the resolvent kernel associated with $k(t)$.

It is now easy to check that $x(t)$ given by (14.8) is a solution of (14.1) for any $h \in L(R, \mathscr{C})$. Since (14.1) and $h \in L(R, \mathscr{C})$ imply (14.8), it follows that $x(t)$ given by (14.8) is the unique solution belonging to $L(R, \mathscr{C})$ of Eq. (14.1).

Let us now prove that (14.8) gives the solution of (14.1) for any $h \in E$, where E stands for any one of the spaces M, S, P_ω, L^p ($1 \le p \le \infty$), C, AP, or A_ω. From Theorem 7.1 we see that (14.8) has a meaning for any $h \in E$ and also that $x \in E$. By direct substitution we shall prove that $x(t)$ given by (14.8) is a solution of (14.1). It suffices to consider the case $E = M(R, \mathscr{C})$. Indeed, we have to prove that

$$h(t) + \int_R k_1(s)h(t - s)\, ds = h(t) + \int_R k(s)[h(t - s)$$

$$+ \int_R k_1(u)h(t - s - u)\, du]\, ds,$$

which leads to

$$\int_R [k_1(s) - k(s) - \int_R k(s - u)k_1(u)\, du]h(t - s)\, ds = 0. \qquad (14.11)$$

But (14.1) is obviously verified if we take into account (14.10). Therefore, Eq. (14.1) has a solution $x \in M$ for any $h \in M$. It remains to show that this solution is unique in M. In other words, we have to prove that (14.1), (14.10), and $x(t) \in M(R, \mathscr{C})$ imply (14.8). This can be done as follows. Let us write (14.1) in the equivalent form

$$x(s) = h(s) + \int_R k(u)x(s - u)\, du, \qquad s \in R, \tag{14.12}$$

and multiply both sides in (14.12) by $k_1(t - s)$. By integration on R with respect to s we obtain

$$\int_R k_1(t - s)x(s)\, ds = \int_R k_1(t - s)h(s)\, ds$$

$$+ \int_R k_1(t - s)\, ds \int_R k(u)x(s - u)\, du. \tag{14.13}$$

If we change s to $v + u$ in the last (double) integral and afterwards again denote v by s, we have

$$\int_R \int_R k_1(t - s)k(u)x(s - u)\, ds\, du = \int_R [k_1(t - s) - k(t - s)]x(s)\, ds, \tag{14.14}$$

if we recall that

$$\int_R k_1(t - s - u)k(u)\, du = k_1(t - s) - k(t - s).$$

The last equation follows from (14.10). From (14.13) and (14.14) we obtain

$$\int_R k_1(t - s)h(s)\, ds - \int_R k(t - s)x(s)\, ds = 0.$$

Taking into account Eq. (14.1) we have

$$\int_R k_1(t - s)h(s)\, ds - x(t) + h(s) = 0. \tag{14.15}$$

But (14.15) is nothing more than (14.8). Therefore, the uniqueness is proved.

As remarked above, $x(t)$ given by (14.8) belongs to the same space E as $h(t)$. Hence, Theorem 14.1 is proved for any space E among the spaces listed in its statement.

Remark 1 If instead of a scalar equation we deal with vector equations, then conditions (14.3) should be replaced by

$$\det(I - \tilde{k}(s)) \neq 0, \qquad s \in R,$$

where I is the unit matrix of the same order as k.

Remark 2   The equation

$$x(t) = h(t) + \int_0^\infty k(t - s)x(s)\, ds, \qquad t \in R_+,$$

needs a much more intricate discussion with respect to the existence of solutions belonging to the corresponding spaces $M(R_+, \mathscr{C})$, $L^p(R_+, \mathscr{C})$, etc. In particular, condition (14.3) does not suffice for the existence of solutions.

Consider now the nonlinear equation (14.2) and let us prove an existence result by means of the contraction mapping principle.

Theorem 14.2   Assume that the following conditions hold for Eq. (14.2):

1. $h(t) \in E$, where E denotes any one of the spaces $M, S, P_\omega, L^p(1 \le p \le \infty)$, C, AP, or A_ω;

2. $k(t)$ is a kernel such that $\|k\| \in L(R, R)$;

3. the mapping $x \to fx$ from $\Sigma = \{x : x \in E,\ |x|_E \le \rho\}$ into E satisfies the Lipschitz condition

$$|fx - fy|_E \le \lambda |x - y|_E; \tag{14.16}$$

4. the inequalities

$$\lambda |k|_L < 1, \qquad |h|_E + |k|_L |f\theta|_E \le \rho(1 - \lambda |k|_L) \tag{14.17}$$

hold, with $|k|_L = \int_R \|k(s)\|\, ds$, θ being the null element of E.

Then there exists a unique solution $x \in \Sigma$ of Eq. (14.2).

Proof   We consider the operator T given by

$$(Tx)(t) = h(t) + \int_R k(s)f(t - s; x)\, ds, \qquad t \in R. \tag{14.18}$$

It is defined on Σ, and we have $T\Sigma \subset \Sigma$ because

$$|Tx|_E \le |h|_E + |k|_L |f(\cdot\,; x)|_E \le |h|_E + |k|_L |fx - f\theta|_E + |k|_L |f\theta|_E$$
$$\le |h|_E + |k|_L |f\theta|_E + \lambda |k|_L |x|_E$$

for any $x \in \Sigma$. That T is a contraction mapping follows easily from the first condition (14.17). The theorem is thereby proved.

Remark 1   An existence theorem can be obtained for Eq. (14.2) by means of the Schauder–Tychonoff fixed-point principle in the case of periodic solutions ($E = P_\omega$ or $E = A_\omega$).

In concluding this section we shall give an existence theorem in C_ℓ. The auxiliary result we need was established in Theorem 7.2.

Theorem 14.3   Consider the equation

$$x(t) = h(t) + \int_0^\infty k(t - s) f(s; x) \, ds, \qquad t \in R_+, \tag{14.19}$$

and assume that the following conditions are satisfied.

1. $h(t) \in C_\ell(R_+, R^n)$;
2. $k(t)$ is a measurable matrix kernel of type n by m such that $\|k\| \in L(R, R)$;
3. the mapping $x \to fx$ from $\Sigma = \{x : x \in C_\ell(R_+, R^n), |x|_C \le \rho\}$ into $C_\ell(R_+, R^m)$ is such that

$$|fx - fy|_C \le \lambda |x - y|_C \tag{14.20}$$

holds for any $x, y \in \Sigma$;

4. the inequalities

$$\lambda |k|_L < 1, \qquad |h|_C + |k|_L |f\theta|_C \le (1 - \lambda |k|_L)\rho \tag{14.21}$$

hold.

Under these assumptions, there exists a unique convergent solution $x \in \Sigma$ of Eq. (14.1).

The proof is straightforward and therefore will be omitted.

2.15   Existence of Measurable Solutions

The admissibility results given in Sections 2.6 and 2.8 will be used in order to obtain some existence theorems for measurable solutions. We shall begin with an existence theorem in the space L_G. The result we shall establish generalizes the existence theorems given in Sections 2.10 and 2.13 in the sense that all continuity requirements are replaced by adequate measurability conditions. In contrast to Section 2.10, in the present section we consider integral operators that are not necessarily of the Volterra type.

Theorem 15.1   Consider the integral equation

$$x(t) = h(t) + \int_R k(t, s) f(s; x) \, ds, \qquad t \in R, \tag{15.1}$$

and assume that the following conditions are satisfied:

1. $h(t) \in L_G^\infty(R, R^n)$, with $G(t)$ a positive measurable function on R;
2. $k(t, s)$ is a measurable matrix kernel of type n by m such that

$$\int_R \|k(t, s)\| g(s) \, ds \le MG(t) \qquad \text{a.e. on } R, \tag{15.2}$$

where M is a positive constant and $g(t)$ is a function with the same properties as $G(t)$;

   3. the mapping $x \to fx$ from $\Sigma = \{x : x \in L_G^\infty(R, R^n), \; |x|_{L_G}{}^\infty \leq \rho\}$ into $L_G^\infty(R, R^m)$ satisfies the Lipschitz condition

$$|fx - fy|_{L_g}{}^\infty \leq \lambda |x - y|_{L_G}{}^\infty; \tag{15.3}$$

   4. the inequalities

$$\lambda M < 1, \qquad |h|_{L_G}{}^\infty + M|f\theta|_{L_g}{}^\infty \leq (1 - \lambda M)\rho \tag{15.4}$$

hold true, with θ representing the null function in $L_G^\infty(R, R^n)$.

Then there exists a unique solution $x \in \Sigma$ of Eq. (15.1).

Proof   Condition (15.2) is the admissibility condition of the pair (L_g^∞, L_G^∞) with respect to the linear integral operator generated by the kernel $k(t, s)$ (see Theorem 6.1). The operator

$$(Tx)(t) = h(t) + \int_R k(t, s)f(s; x)\, ds, \qquad t \in R, \tag{15.5}$$

carries Σ into itself, and under our assumptions it is a contraction. The details of the proof are omitted.

Another existence result is concerned with solutions belonging to the space $L^p, 1 \leq p < \infty$.

Theorem 15.2   Assume that Eq. (15.1) satisfies the following conditions:
   1. $h(t) \in L^p(R, R^n), 1 \leq p < \infty$;
   2. $k(t, s)$ is a measurable matrix kernel of type n by n such that

$$\int_R \|k(t, s)\|g(s)\, ds \in L^p(R, R), \tag{15.6}$$

where $g(t)$ is measurable and positive on R;
   3. the mapping $x \to fx$ from $\Sigma = \{x : x \in L^p(R, \quad R^n), \; |x|_{L^p} \leq \rho\}$ into $L_g(R, R^n)$ satisfies the Lipschitz condition

$$|fx - fy|_{L_g}{}^\infty \leq \lambda |x - y|_{L^p}. \tag{15.7}$$

Then there exists a unique solution $x \in \Sigma$ of Eq. (15.1), as long as $|h|_{L^p}$, $|f\theta|_{L_g}{}^\infty$, and λ are sufficiently small.

Proof   According to Theorem 8.4, condition (15.6) assures the admissibility of the pair (L_g^∞, L^p) with respect to the linear integral operator generated by the kernel $k(t, s)$. The operator T used in the proof of the preceding theorem now acts from Σ into itself, provided $|h|_{L^p}$, $|f\theta|_{L_g}$, and λ are small enough. To be more precise, if M denotes a positive number for which the operator $f \to \int_R k(t, s) f(s)\, ds$, from L_g^∞ into L^p, satisfies

$$\left| \int_R k(t, s) f(s)\, ds \right|_{L^p} \le M |f|_{L_g^\infty},$$

then λ should be such that $\lambda M < 1$, and the following inequality must hold:

$$|h|_{L^p} + M |f\theta|_{L_g^\infty} \le (1 - \lambda M)\rho. \qquad (15.8)$$

This last inequality guarantees the fact that T carries Σ into itself.

Remark 1 When $(fx)(t) = f(t, x(t))$, condition (15.7) can be written in the equivalent form

$$\| f(t, x(t)) - f(t, y(t)) \| \le \lambda g(t) |x - y|_{L^p}.$$

In this case, the following truncation technique opens the way toward a numerical treatment of Eq. (15.1). For any $T > 0$, denote by $x_T(t)$ the function satisfying the "approximate" equation

$$x_T(t) = h(t) + \int_R k(t, s) f(s, x_T)\, ds, \qquad |t| \le T, \qquad (15.9)$$

and vanishing outside the interval $[-T, T]$. Let us remark that Eq. (15.9) can be written as

$$x_T(t) = h_T(t) + \int_{|t| \le T} k(t, s) f(s, x_T)\, ds, \qquad (15.10)$$

where

$$h_T(t) = h(t) + \int_{|t| \le T} k(t, s) f(s, \theta)\, ds. \qquad (15.11)$$

We notice that $f(s, x_T)$ has a meaning because $x_T \in \Sigma$ for $x \in \Sigma$. Moreover, under the assumptions of Theorem 15.2, it follows that $x_T(t)$ is well defined by (15.9) and by the condition that it equals zero outside the interval $[-T, T]$. This also follows from the contraction mapping principle applied to the operator U given by

$$(Ux_T)(t) = h(t) + \int_R k(t, s) f(s, x_T)\, ds, \qquad |t| \le T,$$

$$(Ux_T)(t) = 0 \qquad \text{for} \quad |t| > T.$$

It is our aim to prove that

$$\lim_{T \to \infty} |x - x_T|_{L^p} = 0, \qquad (15.12)$$

which justifies the use of the truncation procedure. We have a.e. on $[-T, T]$

$$\|x(t) - x_T(t)\| = \left\| \int_R k(t, s)[f(s, x) - f(s, x_T)] \, ds \right\|$$

$$\leq \lambda |x - x_T|_{L^p} \int_R \|k(t, s)\| g(s) \, ds,$$

which yields

$$|x - x_T|_{L^p}^p \leq (\lambda M)^p |x - x_T|_{L^p}^p + \int_{|t| \geq T} \|x(s)\|^p \, ds,$$

with

$$M^p = \int_R \left\{ \int_R \|k(t, s)\| g(s) \, ds \right\}^p dt.$$

From the last inequality we obtain

$$|x - x_T|_{L^p} \leq [1 - (\lambda M)^p]^{-1/p} \left\{ \int_{|t| \geq T} \|x(s)\|^p \, ds \right\}^{1/p}$$

and (15.12) is thus proven.

We shall now mention a class of linear integral equations to which the Fredholm theory is applicable. The main requirement for the application of this theory is the complete continuity of the integral operator. Consider the scalar integral equation with parameter

$$x(t) = h(t) + \lambda \int_R k(t, s) x(s) \, ds, \qquad t \in R, \qquad (15.13)$$

and assume that the kernel $k(t, s)$ is a measurable function on $R \times R$ such that

$$\int_R \int_R |k(t, s)|^2 \, dt \, ds < +\infty. \qquad (15.14)$$

As shown in Section 2.8, the linear integral operator $x \to \int_R k(t, s) x(s) \, ds$ is completely continuous from $L^2(R, \mathscr{C})$ into itself. Therefore, the classical results of Fredholm keep their validity for Eq. (15.13), the underlying space being $L^2(R, \mathscr{C})$. For a general Fredholm theory that can be applied to the Eq. (15.13) under condition (15.14), Yosida's book [1] provides the necessary details.

Exercises

1. Let $k(t, s)$ be a continuous matrix kernel for $0 \leq s \leq t < +\infty$ of type m by n, such that $\|k(t, s)\| \leq k_0$. Assume that $g(t)$ is a continuous positive function and $\int_0^\infty g(t)\, dt < +\infty$. Prove that the pair (C_g, C), where $C_g = C_g(R_+, R^n)$, $C = C(R_+, R^m)$, is admissible with respect to the operator K given by $(Kx)(t) = \int_0^t k(t, s)x(s)\, ds$.

2. If $\|k(t, s)\| \leq k_0\, e^{-\alpha(t-s)}$, $\alpha > 0$, and $g(t) \in M(R_+, R)$ is continuous and positive, then the pair (C_g, C) is admissible with respect to the Volterra operator generated by $k(t, s)$.

3. Let $k(t, s)$ be a continuous matrix kernel on $0 \leq s \leq t < +\infty$ of type m by n, also satisfying hypothesis A from Section 2.3. Prove that the pair (C_ℓ, C_ℓ) is admissible with respect to the Volterra operator generated by $k(t, s)$ if and only if the following conditions hold:

1. $\int_0^t k(t, s)\, ds \in C_\ell$, in other words $\int_0^t k_{ij}(t, s)\, ds \in C_\ell(R_+, R)$, $i = 1, 2, \ldots, m$, $j = 1, 2, \ldots, n$;
2. $\int_0^t \|k(t, s)\|\, ds$ is bounded on R_+.

Find $(Kx)(\infty)$ for $x \in C_\ell$.

4. For the same Volterra operator as above, the pair (C_0, C_0) is admissible if and only if condition (2) from Exercise 3 holds and $k(s) \equiv 0$.

5. The pair (C_0, C) is admissible with respect to the Volterra operator whose kernel satisfies hypothesis A if and only if condition (2) from Exercise 3 holds true.

6. A necessary and sufficient condition that the pair (C_0, C_g) be admissible with respect to the Volterra operator (without hypothesis A) generated by the kernel $k(t, s)$ is $\int_0^t \|k(t, s)\|\, ds \in C_g(R_+, R)$.

7. Taking into account the compactness criterion in C_ℓ given in Section 2.12, examine the complete continuity of the Volterra operator occurring in Exercises 3 and 5.

8. Let $g(t) = e^{-\alpha t}$, $t \in R_+$, $\alpha > 0$. If $k(t, s)$ is a measurable matrix kernel of type m by n such that $\|k(t, s)\| \leq Ne^{-\alpha t + \beta s}$ almost everywhere on $R_+ \times R_+$ where $N > 0$ and $\alpha > \beta > 0$, then the pair $(L_g^\infty(R_+, R^n), L_g^\infty(R_+, R^m))$ is admissible with respect to the operator K defined by $(Kx)(t) = \int_0^t k(t, s)x(s)\, ds$.

9. Consider the linear Volterra equation (E): $x(t) = h(t) + \int_0^t k(t, s)x(s)\, ds$. Let us denote by $\gamma(t, s)$ the resolvent kernel associated with $k(t, s)$. We assume that $k(t, s)$ is continuous on $0 \leq s \leq t < +\infty$. A necessary and sufficient condition that Eq. (E) possess a solution $x \in C_G$ for any $h \in C_g$ is $g(t) + \int_0^t \|\gamma(t, s)\|g(s)\, ds \leq MG(t)$, $t \in R_+$, with $M > 0$.

10. Consider the linear differential system

(S) $\dot{x} = A(t)x + u(t)$,   with $\|A(t)\| \in L_{\text{loc}}(R_+, R)$.

Let X_1 be the subspace of R^n consisting of the values at $t = 0$ of all bounded solutions of the homogeneous system $(S_0): \dot{x} = A(t)x$. Let X_2 be any fixed subspace of R^n supplementary to X_1, and let P_1, P_2 denote the corresponding projections of R^n onto X_1, X_2, respectively. If $X(t)$ denotes the fundamental matrix of (S_0) for which $X(0) = I$, then (S) has at least one bounded solution on R_+ for every $u \in L^p(R_+, R^n)$, $1 < p \le \infty$, if and only if there exists $K > 0$ such that

$$\int_0^t \|X(t)P_1 X^{-1}(s)\|^q \, ds + \int_t^\infty \|X(t)P_2 X^{-1}(s)\|^q \, ds \le K^q, \qquad t \in R_+,$$

where $q = p/(p-1)$ for $1 < p < \infty$, $q = 1$ for $p = \infty$. Also discuss the case $p = 1$.

Hint: Observe that

$$x(t) = \int_0^t X(t)P_1 X^{-1}(s)u(s) \, ds - \int_t^\infty X(t)P_2 X^{-1}(s)u(s) \, ds$$

gives a solution of (S).

11. From the preceding exercise derive necessary and sufficient conditions that the pair (L^p, C) be fully admissible with respect to (S).

12. Consider the integro-differential system

(S) $\dot{x} = Ax + \lambda \displaystyle\int_R b(t-s)x(s) \, ds + u(t)$,

where A has characteristic roots with nonzero real parts. Assume further that $\|b\| \in L(R, R)$ and $\|u\| \in L^p(R, R)$, $1 \le p \le \infty$. Then (S) has a unique solution in L^p if $|\lambda|$ is sufficiently small.

13. Let us consider Eq. (E): $x(t) = h(t) + \int_0^t k(t, s)f(s; x) \, ds$, $t \in R_+$, and assume that the following conditions hold:

1. B and D are Banach spaces of functions stronger than $C_c(R_+, R^n)$, such that the pair (B, D) is admissible with respect to the operator $(Kf)(t) = \int_0^t k(t, s)f(s) \, ds$;

2. the mapping $x \to fx$ carries the ball $\Sigma = \{x: x \in D, \ |x|_D \le \rho\}$ into B and the Lipschitz condition $|fx - fy|_B \le \lambda |x - y|_D$ is verified;

3. $h \in D$.

Then there exists a unique solution $x \in \Sigma$ of Eq. (E) as long as

$$\lambda < \|K\|^{-1}, \qquad |h|_D + \|K\| \, |f\theta|_B \le (1 - \lambda\|K\|)\rho.$$

14. Consider the integral equation (E) from Exercise 13 and assume that the following conditions hold:

1. and 2.—the same as in Exercise 3;
3. the mapping $x \to fx$ from $C_\ell(R_+, R^n)$ into itself satisfies the Lipschitz condition $|fx - fy|_C \le \lambda|x - y|_C$;
4. $h \in C_\ell(R_+, R^n)$.

Then there exists a unique solution $x \in C_\ell(R_+, R^n)$ of Eq. (E) as long as λ is sufficiently small.

15. State and prove an existence theorem for Eq. (E), similar to that given in Exercise 14, using the fixed-point theorem of Schauder and Tychonoff.

16. Let us again consider Eq. (E) from Exercise 13 and assume that the following conditions hold:

1. $k(t, s)$ is continuous for $0 \le s \le t < +\infty$ and satisfies hypothesis A from Section 2.3 with $k(s) \equiv 0$;
2. $\int_0^t \|k(t, s)\| g(s) \, ds \in C_0(R_+, R)$, where $g(t)$ denotes a positive continuous function on R_+;
3. the mapping $x \to fx$ from $\Sigma = \{x : x \in C_0(R_+, R^n), |x|_C \le \rho\}$ into $C_g(R_+, R^m)$ satisfies the Lipschitz condition $|fx - fy|_{C_g} \le \lambda|x - y|_C$;
4. $h(t) \in C_0(R_+, R^n)$.

Then there exists a unique solution $x \in \Sigma$ of Eq. (E), provided λ, $|h|_C$, and $|f\theta|_{C_g}$ are sufficiently small.

17. Consider the convolution equation

$$(E) \qquad x(t) = h(t) + \int_R k(t - s) f(s; x) \, ds, \qquad t \in R,$$

and let F be any one of the spaces P_ω, A_ω. If $k \in L(R, R)$ and if the mapping $x \to fx$ is continuous from $\Sigma = \{x : x \in F, |x|_F \le \rho\}$ into F and carries Σ into a bounded set, then there exists at least one solution $x \in \Sigma$ of Eq. (E), provided $|h|_F + |k|_L|fx|_F \le \rho$ for any $x \in \Sigma$.

18. Consider Eq. (E): $x(t) = h(t; x) + \int_R k(s)x(t - s) \, ds$, $t \in R$, and assume that the following conditions are fulfilled:

1. $k(t) \in L(R, \mathscr{C})$ and $\tilde{k}(s) \ne 1$ for $s \in R$, where $\tilde{k}(s) = \int_R k(t)e^{its} \, dt$;
2. the mapping $x \to hx$ is Lipschitzian from $\Sigma = \{x : x \in F, |x|_F \le \rho\}$ into F, where F denotes any one of the spaces M, S, or P_ω: $|hx - hy|_F \le \lambda|x - y|_F$;
3. the following inequalities hold:

$$\lambda < (1 + |k_1|_L)^{-1}, \qquad |h\theta|_F \le [(1 + |k_1|_L)^{-1} - \lambda]\rho,$$

where $k_1(t)$ is the resolvent kernel associated with $k(t)$.

Then (E) has a unique solution $x \in \Sigma$.

19. Consider the integral equation (E): $x(t) = h(t) + \int_R k(s)f(t - s; x)\, ds$, $t \in R$, and assume that:

1. $h(t), k(t) \in M(R, R)$;
2. the mapping $x \to fx$ from $\Sigma = \{x : x \in M, |x|_M \le \rho\}$ into $L(R, R)$ satisfies the Lipschitz condition $|fx - fy|_L \le \lambda |x - y|_M$;
3. the inequalities

$$\lambda |k|_M < 1, \qquad |h|_M + |k|_M |f\theta|_L \le (1 - \lambda |k|_M)$$

are verified.

Then there exists a unique solution $x \in \Sigma$ of Eq. (E). Discuss the special cases when h and k belong to S or P_ω. In the last situation prove an existence result by means of the Schauder–Tychonoff fixed-point theorem.

20. Find examples of operators acting from M into L. Use the result given in Exercise 2, Section 1.1.

21. Consider the integral equation (E): $x(t) = h(t; x) + \int_R k(t, s)f(s; x)\, ds$. Discuss existence and uniqueness in various function spaces, using the results and methods given in Chapter II.

Bibliographical Notes

The main result of Section 2.2 is due to the author [3]. A similar result was proved by Youla [1]. The admissibility theorem of Section 2.3 is due to Bantaş [1]. The results given in Sections 2.4 and 2.5 are essentially due to Radon (see Zabreiko *et al.* [1]). See also the author's paper [9]. For the proof of Theorem 6.1, the author is indebted to Foiaş and Zabreiko. See also Petrovanu's paper [1]. The admissibility results related to the convolution operator are classical. The case of the space M seems to be discussed for the first time in the author's paper [11].

The significance of the admissibility results for the theory of linear (engineering) systems was illustrated in a polemic discussion waged by Kalman [1] and Bridgland [1].

In his recent paper O'Neil [1] found many interesting results related to the admissibility theory for integral operators. The case of Orlicz spaces is discussed.

The results of Sections 2.10 and 2.11 are due to the author [3, 5, 6]. Those included in Section 2.12 are due to Bantaş [1], except for Lemma 12.1 and Theorem 12.3, which were established by the author. The results of Section 2.13 are published for the first time in this book, and they were found by the author. Some of the results included in Sections 2.14 are classical. Some of them can be found in the author's papers [6, 10]. The results of Section 2.15 are due to Petrovanu [1].

The mathematical literature provides a good deal of information concerning the existence, uniqueness, and continuous dependence of solutions for various classes of Hammerstein equations. See Krasnoselski [1], Lakshmikantham and Leela [1], Nohel [1–4], and Zabreiko *et al.* [1]. Here we shall discuss some results related to those expounded in this chapter.

In a series of papers dedicated to the theory of stability of Volterra integral equations, Vinokurov [1–3] found many interesting results related to the behavior on R_+ of solutions of integral equations of the form

(E) $$x(t) = h(t) + \int_0^t k(t, s)x(s)\, ds$$

or

(E′) $$x(t) = h(t) + \int_0^t k(t, s, x(s))x(s)\, ds.$$

Let us give some details concerning the linear equation (E). We shall consider some concepts of stability for (E).

The solution of (E) is called stable if to any $\varepsilon > 0$, there corresponds $\delta = \delta(\varepsilon) > 0$ such that $\|h(t)\| < \delta$ on R_+ implies $\|x(t)\| < \varepsilon$ on R_+. If $\gamma(t, s)$ is the associated resolvent kernel for $k(t, s)$ then a necessary and sufficient condition for stability is $\int_0^t \|\gamma(t, s)\|\, ds \le M$ for $t \in R_+$.

The solution of (E) is called asymptotically stable if to any $\varepsilon > 0$, $\mu > 0$, there correspond $\delta_1(\varepsilon) > 0$, $\delta_2(\varepsilon) > 0$, $\delta_3(\mu) > 0$, and $T(\mu) \ge 0$, such that $\|h(0)\| < \delta_1$, $\|h'(t)\| < \delta_2$, on R_+, and $\|h'(t)\| < \delta_3$ on $[T, \infty)$ imply $\|x(t)\| < \varepsilon$ on R_+ and $\|x(t)\| < \mu$ on $[T, \infty)$. A necessary and sufficient condition for asymptotic stability is $\lim_{t \to \infty}(I + \int_0^t \gamma(t, s)\, ds) = 0$, where 0 denotes the null matrix and I, the unit matrix.

Using the same concept of stability as above and the theory of integral inequalities of the Volterra type, Calyuk [1] found further stability results and also considered integral equations with delay.

Goldenhershel [1] investigated the behavior of solutions of a Volterra equation using some spectral properties of the linear integral operator $(Tx)(t) = \int_0^t k(t, s)x(s)\, ds$. The estimate of the spectral radius provides a useful tool in investigating the equation $Tx - \lambda x = f$, with f a fixed element. Interesting results were found in connection with the exponential growth of solutions. Similar procedures were used by A. Corduneanu [1–3] in the case of L^p spaces. It is shown in [1] that from $\{\int_0^x |k(t, s)|^q\, ds\}^{1/q} \in L^p([x, \infty), R)$ for any $x > 0$, $p^{-1} + q^{-1} = 1$, there results the following estimate for the spectral radius of T in L^p.

$$r(T) \le \lim_{t \to \infty}\left\{\int_t^\infty dx\left[\int_t^x |k(x, y)|^q\, dy\right]^{p/q}\right\}^{1/p}.$$

The spectrum enjoys many other useful properties in studying integral equations.

Bihari [1] considered the integral equation

$$(E) \qquad x(t) = z(t) + \int_0^t k_1(t, s) f(s, x(s)) \, ds$$

$$+ \int_t^\infty k_2(t, s) f(s, x(s)) \, ds, \qquad t \in R_+, \qquad x, z \in R^n.$$

The following existence result was established. Assume that (E) satisfies these conditions:

1. $k_1(t, s)$ and $k_2(t, s)$ are continuous and bounded in $0 \leq s \leq t < +\infty$ and $0 \leq t \leq s < +\infty$, respectively, i.e., $\|k_i(t, s)\| \leq K_i$, $i = 1, 2$;

2. $f(t, x)$ is continuous from $R_+ \times R^n$ into R^n and $\|f(t, x)\| \leq G(t, \|x\|)$, where $G(t, r)$ is piecewise continuous on $R_+ \times R_+$ and nondecreasing (for fixed t) with respect to r;

3. the inequality

$$\gamma + K_1 \int_0^t G(s, g(s)) \, ds + K_2 \int_t^\infty G(s, g(s)) \, ds \leq g(t)$$

holds for a certain $g(t)$, positive and continuous on R_+, with $\gamma > 0$ an arbitrary constant;

4. $z(t) \in C(R_+, R^n)$.

Then there exists at least one solution $x(t)$ of Eq. (E) such that $x \in C_g(R_+, R^n)$ and $|x|_{C_g} \leq 1$.

Various results concerning the existence of solutions for Hammerstein equations with symmetric kernel can be found in Krasnoselski's book [1].

Since any differential system of the form $\dot{x} = A(t)x + f(t, x)$, $t \in R_+$, can be transformed into an integral equation of Hammerstein type (with a Volterra operator or, more generally, with an operator of the form $f \to \int_0^\infty k(t, s) f(s) \, ds$, the results established in this chapter can be easily applied. For a treatment of this kind of problem we recommend the book by Hartman [1]. Further references can be found in this book and also in Coppel's book [1]. The papers of Conti [1] and Reghiş [1] contain interesting results related to this subject. We want to point out that Bellman [1] was the first author to make clear the significance of the admissibility conditions in studying the asymptotic properties of solutions of differential systems.

A noteworthy contribution related to the theory of differential systems of the form $\dot{x} = A(t)x + f(t, x)$, $t \in R_+$, $x \in H = $ a Hilbert space, was given by

Browder [1]. Let us mention a result from the paper quoted above. Assume that $\{A(t): t \in R_+\}$ is a family of closed linear operators such that the following assumptions are verified: (a) the domain of definition $D(A(t))$ is independent of t for $t \in R_+$; (b) the mapping $t \to A(t)$ from R_+ into the space of operators from D to H, with the strong operator topology, is of class $C^{(1)}$ (the topology of D being that given by the graph norm of $A(0)$); (c) if $A^*(t)$ is the adjoint of $A(t)$, then $D(A^*(t)) \subset D(A(t))$; (d) for each $N > 0$, there exists a constant c_N such that $\mathrm{Re}(A(t)u, u) \le c_N \|u\|_H^2$, for all $u \in D(A(t))$ and $t \in [0, N]$; (e) f is a continuous mapping from $R_+ \times H$ into H carrying bounded sets into bounded sets, and, for each $N > 0$, there exists a constant γ_N such that $\mathrm{Re}(f(t, u) - f(t, v), u - v) \le \gamma_N \|u - v\|_H^2$, for all $u, v \in H$ and $t \in [0, N]$. Then there exists at least one (generalized) solution of the given system, defined on R_+ and satisfying an initial value condition $x(0) = x_0 \in H$. More precisely, this solution satisfies the integral equation of Hammerstein type $x(t) = X(t, 0)x_0 + \int_0^t X(t, s) f(s, x(s)) \, ds$, $t \in R_+$, where $X(t, s)$ is the transition operator for $\dot{x} = A(t)x$ (i.e., $X(t, s)\tilde{x} = x(t)$ is the solution of $\dot{x} = A(t)x$ with $x(s) = \tilde{x}$). The above result is obtained by means of a theorem concerning the range of a monotone nonlinear operator in a Hilbert space. We take the opportunity to point out that a good deal of the existence theorems for integral equations could be obtained using the theory of monotone operators (see the quoted paper by Browder for extensions and references).

Several results are known with respect to nonlinear integral equations of the form (E): $x(t) = h(t) + \int_0^t k(t, s, x(s)) \, ds$, $t \in R_+$. We shall mention here a result obtained by means of Schauder–Tychonoff fixed-point principle. Assume that the following conditions hold for (E): (a) the mapping $(t, s, x) \to k(t, s, x)$ is continuous from $D = \{(t, s, x): 0 \le s \le t < +\infty, x \in R^n, \|x\| \le g(t)\}$ into R^n, where $g(t)$ is a continuous positive function on R_+; (b) there exists a continuous function $k(t, s)$, $0 \le s \le t < +\infty$, such that $\|k(t, s, x)\| \le k(t, s)$ for $(t, s, x) \in D$; (c) $h(t) \in C_c(R_+, R^n)$; (d) $\|h(t)\| + \int_0^t k(t, s) \, ds \le g(t)$, $t \in R_+$. Then there exists at least one solution $x \in C_g(R_+, R^n)$ of Eq. (E) such that $|x|_{C_g} \le 1$. In order to derive existence results for Hammerstein equations we have to take $k(t, s, x) = k(t, s)f(s, x)$ in (E). For details see the author's paper [3].

A detailed discussion of the existence, uniqueness, and continuous dependence on the data of solutions of integral equations of the form $x(t) = h(t) + \int_0^t k(t, s)f(s, x(s)) \, ds$, under measurability assumptions, can be found in the paper by Miller and Sell [1].

Results related to those given in this chapter can be also found in the book by Lakshmikantham and Leela [1]. The method of comparison (integral inequalities) is particularly emphasized in this book. Further results on Volterra integral equations and useful references are given by Nohel in his papers [1–4].

The author's survey paper [10] discusses some topics related to the admissibility theory with respect to an integral operator and the theory of Hammerstein equations. The application of the admissibility techniques to the theory of stochastic integral equations was made by Tsokos in [1]. It seems that further results in this field could be obtained by developing the theory of integral equations in Hilbert or Banach spaces. Very interesting contributions concerning the admissibility theory for integral operators have been made recently by Gollwitzer [1].

Frequency Techniques and Stability

This chapter is devoted to investigation of some nonlinear convolution equations—mainly of Volterra type—with special emphasis on stability results. The theory due to Popov, as well as its extensions and applications, constitutes the core of the present chapter.

It is well known that an important problem occurring in the study of various classes of physical systems is the determination of conditions under which such a system is stable. Usually, by stability is meant a property of the system consisting of the fact that the response to a certain class of inputs approaches zero as $t \to \infty$. Several generalizations of this classical concept are discussed in the literature and they will find a place in our exposition.

Since many physical systems are governed by nonlinear integral equations of convolution type, the results we are going to establish have many interesting applications. First, we mention the stability problem of the automatic control system. The methods we shall use in studying the behavior of solutions of integral equations apply with minor changes to the investigation of various classes of nonlinear feedback systems.

3.1 General Remarks and Statement of Some Problems

Let us consider the nonlinear integral equation of convolution type

$$\sigma(t) = h(t) + \int_0^t k(t-s)\varphi(\sigma(s))\, ds, \qquad t \in R_+, \tag{1.1}$$

where σ, h, k, and φ are scalar functions. As pointed out in the preface, Eq. (1.1) can be interpreted as describing a feedback system that contains a linear part whose input–output equation is

$$\sigma(t) = h(t) + \int_0^t k(t-s)u(s)\, ds, \qquad t \in R_+, \tag{1.2}$$

and a nonlinear element, the input–output equation of which is

$$u(t) = \varphi(\sigma(t)), \qquad t \in R_+. \tag{1.3}$$

In other words, our system is a closed feedback loop. One of the possible interpretations of this system is: $\sigma(t)$—the signal in the feedback loop; $h(t)$—a signal applied externally to the system or formed as a result of certain initial conditions; $k(t)$—the pulse response of the linear part; φ—the characteristic of the nonlinear element occurring in the system.

It is interesting to derive Eq. (1.1) starting from the systems of automatic control described by differential equations (see Aizerman and Gantmacher [1]; Lefschetz [1]). This will suggest to us the kind of hypotheses to be used in order to get results with interesting applications.

The system of equations corresponding to a direct control problem is

$$\dot{x} = Ax + b\varphi(\sigma), \qquad \sigma = (c, x), \tag{1.4}$$

where A is a constant matrix, b and c are constant vectors, φ is the characteristic function of the servomotor, and x is an unknown state vector. From the first equation in (1.4) we obtain

$$x(t) = e^{At}x^0 + \int_0^t e^{A(t-s)}b\varphi(\sigma(s))\, ds,$$

with x^0 an arbitrary vector. If we substitute the above expression for x in the second equation in (1.4), we obtain

$$\sigma(t) = (c, e^{At}x^0) + \int_0^t (c, e^{A(t-s)}b)\varphi(\sigma(s))\, ds, \tag{1.5}$$

which is an equation of the form (1.1) with

$$h(t) = (c, e^{At}x^0), \tag{1.6}$$

$$k(t) = (c, e^{At}b). \tag{1.7}$$

If we assume that A is stable, it follows then that h and k have exponential decay at infinity.

A system of indirect control is described by the equations

$$\dot{x} = Ax + b\xi, \qquad \dot{\xi} = \varphi(\sigma), \qquad \sigma = (c, x) - \rho\xi, \tag{1.8}$$

where x, A, b, and c have the same meaning as above, and ρ is a constant (scalar). We obtain

$$x(t) = e^{At}x^0 + \int_0^t e^{A(t-s)}b\xi(s)\,ds,$$

$$\xi(t) = \xi_0 + \int_0^t \varphi(\sigma(s))\,ds,$$

which yield for σ an integral equation of the form (1.1) with

$$h(t) = (c, e^{At}(x^0 + A^{-1}b\xi_0)) - \rho_1\xi_0, \tag{1.9}$$

$$k(t) = (c, A^{-1}e^{At}b) - \rho_1, \tag{1.10}$$

where $\rho_1 = \rho + (c, A^{-1}b)$. From (1.9) and (1.10) one sees that for any stable matrix A, both h and k can be represented as the sum of a constant and a function that decays exponentially at infinity.

If the system of direct control (1.4) is such that the matrix A has zero as a simple characteristic root and the remaining characteristic roots have negative real parts, then by a suitable change of variables (see Lefschetz [1]), (1.4) can be written in the form (1.8). Therefore, there is no need to differentiate between direct controls, corresponding to the critical case when A has zero as a simple characteristic root, and indirect controls, as described by (1.8), with A stable.

The situation becomes different when we deal with other critical cases of direct control. For instance, if the matrix A in (1.4) has zero as a double characteristic root, then (1.4) can be transformed into an equivalent system of the form

$$\dot{x} = Bx + b\varphi(\sigma), \qquad \dot{\eta} = \xi, \qquad \dot{\xi} = \varphi(\sigma), \qquad \sigma = (c, x) - d\xi - e\eta,$$

where B is a stable matrix. Applying the same procedure of reduction to an integral equation, we get an equation of the form (1.1) with

$$h(t) = (c, e^{Bt}x^0) - (d\xi_0 + e\eta_0) - e\xi_0 t,$$

$$k(t) = (c, e^{Bt}b) - d - et.$$

Therefore, both h and k are representable as a sum of a term with exponential decay at infinity and a linear polynomial in t. Another critical case corresponds to direct control systems for which A has a pair of conjugate pure imaginary characteristic roots. In this case, both h and k can be represented as a sum of a term vanishing exponentially at infinity and a trigonometric polynomial of the form $a \cos \omega t + b \sin \omega t$.

Our basic problem in this chapter consists in finding suitable conditions concerning the kernel k of Eq. (1.1), under which $\sigma(t) \in C_0(R_+, R)$ $[\sigma(t) \in C_0(R_+, R^n)$ in the vector case] no matter how we choose the free term h and the function φ within certain classes of functions. Usually, the frequency technique we shall develop leads to some conditions regarding the Fourier transform of the kernel or, in certain cases, the Fourier transform of one sum and of the kernel. These conditions require that certain functions be positive on the real axis.

The kind of behavior characterized by the inclusion $\sigma(t) \in C_0(R_+, R)$ is, of course, the classical stability property. Several extensions are treated in the literature; we shall deal mainly with the so-called energetic stability (see Kudrewicz [1], [2]). Instead of the condition $\sigma(t) \in C_0(R_+, R)$, it is required that

$$\lim_{T \to \infty} T^{-1} \int_0^T |\sigma(s)|^2 \, ds = 0. \tag{1.11}$$

It is obvious that (1.11) represents a weaker condition on σ than

$$\sigma(t) \in C_0(R_+, R).$$

Another interesting problem is that of existence of solutions whose behavior can be successfully investigated by using frequency techniques. In the case of the control system described by differential equations, the existence problem is quite easy. That of the behavior of solutions requires a more intricate discussion. We shall see that the existence problem for various classes of integral equations can also be solved, under reasonable general conditions. It is worth pointing out that the existence problem can be solved more easily in certain function spaces, usually those richer than $C_0(R_+, R)$. Afterward, the frequency method allows us to prove that the solution belongs to $C_0(R_+, R)$. Sometimes the frequency condition itself is used in the proof of the existence result.

3.2   Equation (1.1) with an Integrable Kernel

We shall discuss in this section the equation

$$\sigma(t) = h(t) + \int_0^t k(t-s)\varphi(\sigma(s)) \, ds, \qquad t \in R_+, \tag{1.1}$$

under the main assumption that the kernel $k(t)$ is integrable on R_+:

$$k \in L(R_+, R). \tag{2.1}$$

Theorem 2.1 Consider Eq. (1.1) under the following assumptions:

1. $h(t) \in C(R_+, R)$;
2. $k(t)$ satisfies condition (2.1);
3. φ is continuous and bounded from R into itself, i.e., $\varphi \in C(R, R)$.

Then there exists at least one solution $\sigma(t) \in C(R_+, R)$ of Eq. (1.1).

Proof The operator T given by

$$(T\sigma)(t) = h(t) + \int_0^t k(t - s)\varphi(\sigma(s))\, ds, \qquad t \in R_+, \tag{2.2}$$

is defined on $C_c(R_+, R)$, and takes its values in the same space. It is obviously a continuous operator in the topology of the space $C_c(R_+, R)$. Moreover, if we denote by ϕ a positive number such that

$$|\varphi(\sigma)| \le \phi, \qquad \sigma \in R, \tag{2.3}$$

then we get from (2.3):

$$|(T\sigma)(t)| \le |h|_c + \phi|k|_L, \qquad \sigma(t) \in C_c(R_+, R). \tag{2.4}$$

Let us now consider the ball

$$\sum = \{\sigma(t) : \sigma \in C(R_+, R), |\sigma|_c \le r\}, \tag{2.5}$$

where $r = |h|_c + \phi|k|_L$. It follows then from (2.4) that $T\Sigma \subset \Sigma$. In order to prove the existence of a fixed point for the operator T by means of the Schauder–Tychonoff theorem, it suffices to show that $T\Sigma$ is relatively compact in the topology of the space $C_c(R_+, R)$. This follows easily by applying the classical criterion of compactness for families of continuous functions on a compact interval. The theorem is thus proven.

Remark 1 The same method of proof applies to the more general equation

$$\sigma(t) = h(t) + \int_0^\infty k(t - s)\varphi(\sigma(s))\, ds, \tag{2.6}$$

under condition (2.1) for the kernel.

It is obvious that the proof remains valid in the vector case.

Remark 2 The boundedness assumption (2.3) can be weakened, still keeping the method of proof. Indeed, if instead of (2.3) we assume that to any $M > 0$, there corresponds $\phi(M) > 0$ such that

$$(|\sigma| \le M + |k|_L \phi(M)) \Rightarrow (|\varphi(\sigma)| \le \phi(M)), \tag{2.7}$$

then the ball Σ defined by (2.5) is carried into itself by the operator T, if we take $r = M + |k|_L \phi(M)$, M being such that $|h(t)| \le M$, $t \in R_+$.

We are now able to prove a theorem that gives sufficient conditions for the existence of a solution of Eq. (1.1) in the space $C_0(R_+, R)$.

Theorem 2.2   Assume that the following conditions are satisfied by Eq. (1.1):

1. $h(t)$, $h'(t) \in L(R_+, R)$;
2. $k(t)$, $k'(t) \in L(R_+, R)$;
3. the function $\varphi(\sigma)$ is continuous and bounded from R into itself and satisfies

$$\sigma\varphi(\sigma) > 0 \qquad \text{for } \sigma \neq 0; \tag{2.8}$$

4. there exists $q \geq 0$ such that

$$\text{Re}\{(1 - isq)\tilde{k}(s)\} \leq 0, \qquad s \in R, \tag{2.9}$$

where $\tilde{k}(s)$ denotes the Fourier transform of $k(t)$.

Then there exists at least one solution $\sigma(t) \in C_0(R_+, R)$ of Eq. (1.1). Moreover, any solution $\sigma(t) \in C_c(R_+, R)$ of this equation belongs to $C_0(R_+, R)$.

Proof   Let us first remark that the conditions of our theorem imply those of Theorem 2.1. Indeed, from $h(t)$, $h'(t) \in L(R_+, R)$, it follows that $h(t)$ is continuous and bounded on R_+. Conditions (2) and (3) of Theorem 2.1 are obviously satisfied if we assume that the conditions of Theorem 2.2 hold. Consequently, Eq. (1.1) has at least one bounded (continuous) solution on R_+. Moreover, if $\sigma(t) \in C_c(R_+, R)$ is a solution of Eq. (1.1), then it is necessarily bounded on R_+ because the right member of this equation is always bounded by $|h|_C + |k|_L \phi$, the number ϕ being such that (2.3) holds.

We shall prove now that any solution $\sigma(t)$ of (1.1), continuous on R_+, approaches zero as $t \to \infty$. In other words, $\sigma(t) \in C_0(R_+, R)$. At this point the frequency method will be applied. Let $\sigma(t)$ be a solution of (1.1) continuous on R_+. For any $t > 0$, let us define $\varphi_t(\tau)$ by

$$\varphi_t(\tau) = \begin{cases} \varphi(\sigma(\tau)), & 0 \leq \tau \leq t, \\ 0, & \tau > t. \end{cases} \tag{2.10}$$

Consider another auxiliary function given by

$$\lambda_t(\tau) = \int_0^\tau [k(\tau - u) + qk'(\tau - u)]\varphi_t(u)\, du + qk(0)\varphi_t(\tau) \tag{2.11}$$

for any $\tau \in R_+$. From Eq. (1.1) we obtain by differentiation

$$\sigma'(t) = h'(t) + k(0)\varphi(\sigma(t)) + \int_0^t k'(t - s)\varphi(\sigma(s))\, ds, \tag{2.12}$$

which holds almost everywhere on R_+. Using (1.1) and (2.12), it is possible to give another form to $\lambda_t(\tau)$. Namely,

$$\lambda_t(\tau) = \begin{cases} \sigma(\tau) + q\sigma'(\tau) - [h(\tau) + qh'(\tau)], & 0 \leq \tau \leq t \\ \int_0^t [k(\tau - u) + qk'(\tau - u)]\varphi(\sigma(u))\, du, & \tau > t, \end{cases} \quad (2.13)$$

taking into account that

$$\int_0^\tau k(\tau - u)\varphi(\sigma(u))\, du = \sigma(\tau) - h(\tau),$$

and

$$\int_0^\tau k'(\tau - u)\varphi(\sigma(u))\, du = \sigma'(\tau) - h'(\tau) - k(0)\varphi(\sigma(\tau)).$$

It can be easily seen that $\lambda_t \in L \cap L^2$ on R_+.

If we denote by $\tilde{\varphi}_t(s)$ and $\tilde{\lambda}_t(s)$ the Fourier transforms of the functions $\varphi_t(\tau)$ and $\lambda_t(\tau)$, respectively, then

$$\tilde{\lambda}_t(s) = (1 - isq)\tilde{k}(s)\tilde{\varphi}_t(s). \quad (2.14)$$

Indeed, from (2.11) we see that $\lambda_t(\tau)$ involves the convolution product of $k + qk'$ and φ_t. But the Fourier transform of k' is $-is\tilde{k}(s) - k(0)$, because

$$\int_0^\infty k'(t)e^{ist}\, dt = \int_0^\infty e^{ist}\, dk(t) = e^{ist}k(t)\Big|_0^\infty - is\int_0^\infty k(t)e^{ist}\, dt$$

and $k(\infty) = 0$. The last statement follows from the fact that $k(\infty)$ exists, since $k' \in L(R_+, R)$. From $k \in L(R_+, R)$, there results $k(\infty) = 0$.

Now, we consider another auxiliary function $\rho(t)$ defined by

$$\rho(t) = \int_0^t \lambda_t(\tau)\varphi(\sigma(\tau))\, d\tau = \int_0^\infty \lambda_t(\tau)\varphi_t(\tau)\, d\tau. \quad (2.15)$$

Parseval's equality (see Exercise 1, Section 1.3) applies and we obtain

$$\rho(t) = (2\pi)^{-1}\int_R \tilde{\lambda}_t(s)\overline{\tilde{\varphi}_t(s)}\, ds. \quad (2.16)$$

Taking into account (2.14), we can write further

$$\rho(t) = (2\pi)^{-1}\int_R (1 - isq)\tilde{k}(s)|\tilde{\varphi}_t(s)|^2\, ds. \quad (2.17)$$

By definition, $\rho(t)$ is a real-valued function. Therefore (2.17) yields

$$\rho(t) = (2\pi)^{-1}\int_R \mathrm{Re}\{(1 - isq)\tilde{k}(s)\}|\tilde{\varphi}_t(s)|^2\, ds. \quad (2.18)$$

From (2.9) there results that $\rho(t) \leq 0$ for $t \in R_+$, which means that

$$\int_0^t \varphi(\sigma(\tau))\sigma(\tau) \, d\tau + q \int_0^t \varphi(\sigma(\tau))\sigma'(\tau) \, d\tau - \int_0^t [h(\tau) + qh'(\tau)]\varphi(\sigma(\tau)) \, d\tau \leq 0$$

for every $t \in R_+$. To obtain the last inequality from $\rho(t) \leq 0$, it suffices to observe (2.13). We get further

$$\int_0^t \varphi(\sigma(\tau))\sigma(\tau) \, d\tau + q \int_{\sigma(0)}^{\sigma(t)} \varphi(\sigma) \, d\sigma \leq \phi \int_0^\infty [\, |h(\tau)| + q\,|h'(\tau)|\,] \, d\tau,$$

the number ϕ being chosen such that (2.3) holds. But the second term in the left-hand side of the above inequality is always positive, according to (2.8). Therefore, we can write

$$\int_0^t \varphi(\sigma(\tau))\sigma(\tau) \, d\tau \leq A, \qquad t \in R_+, \tag{2.19}$$

A being a positive constant that depends on the equation. Let us remark now that the integrand in (2.19) is nonnegative and uniformly continuous on R_+. The first property is a consequence of (2.8), while the second one follows easily from (2.12). Indeed, (2.12) shows that $\sigma'(t)$ is dominated by a constant plus an integrable function on R_+, and this implies that $\sigma(t)$ is uniformly continuous.

It remains to apply the following lemma of Barbălat [1]: Let $f(t) \geq 0$ be integrable on R_+ and uniformly continuous. Then $f(t) \to 0$ as $t \to \infty$. One obtains from (2.19) that $\varphi(\sigma(t))\sigma(t) \to 0$ as $t \to \infty$, which takes place if and only if $\sigma(t) \to 0$ as $t \to \infty$ (see (2.8)). Therefore $\sigma(t) \in C_0(R_+, R)$, and the theorem is thus proved.

For the sake of completeness, let us sketch the proof of Barbălat's lemma. Assume that $f(t)$ does not approach zero as $t \to \infty$. Then there results the existence of a sequence $\{t_n\}$, $t_n \to \infty$, such that $f(t_n) > \alpha > 0$ for any $n \geq 1$. The uniform continuity of f assures the existence of a positive β with the property that $f(t) > \frac{1}{2}\alpha$ for $|t - t_n| \leq \beta$, $n \geq 1$. Without loss of generality we can assume that the intervals $(t_n - \beta, t_n + \beta)$ do not overlap. Therefore

$$\int_0^\infty f(t) \, dt \geq \sum_{n=1}^N \int_{t_n - \beta}^{t_n + \beta} f(t) \, dt \geq N\alpha\beta$$

for any N, which contradicts the assumption that f is integrable on R_+. Let us remark that this lemma will be repeatedly applied in the subsequent sections.

In concluding this section, we shall apply the result established in Theorem 2.2 to the integral equation (1.5). As shown in the preceding section, the differential system (1.4) yields Eq. (1.5). We assume that A is a stable matrix and also that $\varphi(\sigma)$ satisfies the conditions specified in the statement of

Theorem 2.2. Then $h(t)$ given by (1.6) goes exponentially to zero as $t \to \infty$. The same is true for the derivative $h'(t)$, and this shows that condition (1) of Theorem 2.2 is fulfilled. A similar argument shows that $k(t)$ satisfies condition (2) of Theorem 2.2. It remains therefore to give the last form to the frequency condition (2.9), taking into account formula (1.7) for $k(t)$. We can write

$$A \int_0^\infty e^{At} e^{ist} \, dt = \int_0^\infty e^{ist} \, de^{At} = e^{ist} e^{At} \Big|_0^\infty - is \int_0^\infty e^{At} e^{ist} \, dt$$

$$= -I - is \int_0^\infty e^{At} e^{ist} \, dt,$$

from which we get

$$\int_0^\infty e^{At} e^{ist} \, dt = -(isI + A)^{-1}, \qquad s \in R. \tag{2.20}$$

Let us remark that the matrix $isI + A$ is nonsingular for any $s \in R$, because $\det(A + isI) = 0$ would imply that A has characteristic roots with zero real part (contrary to the assumption that A is stable). Condition (2.9) can now be written in the form

$$\text{Re}\{(isq - 1)(c, (isI + A)^{-1} b)\} \le 0, \qquad s \in R. \tag{2.21}$$

From (2.21) there results that the solution $\sigma(t)$ of (1.5) approaches zero as $t \to \infty$ (the existence of at least one solution follows from Theorem 2.1). If we take into account the formula

$$x(t) = e^{At} x^0 + \int_0^t e^{A(t-s)} b\varphi(s) \, ds,$$

we obtain that any solution $x(t)$ of the system (1.4) tends to zero as $t \to \infty$.

Consequently, any solution of the system (1.4), with stable A and $\varphi(\sigma)$ satisfying condition (3) from Theorem 2.2, tends to zero as $t \to \infty$, provided that (2.21) holds true. This stability property of the system (1.4) is called absolute stability (see Lefschetz [1]).

3.3   Equation (1.1) with $k(t) + \rho$ Integrable, $\rho > 0$

We shall discuss in this section Eq. (1.1) when its kernel can be written in the form

$$k(t) = k_0(t) - \rho, \qquad k_0 \in L(R_+, R), \tag{3.1}$$

$\rho > 0$ being a given number. As seen in Section 3.1, such a kernel appears when reducing the system (1.4) to an integral equation, under the assumption that A has zero as a simple characteristic root.

The following result will be proved below.

Theorem 3.1   Consider the integral equation (3.1) under the following assumptions:

1. $h'(t)$, $h''(t) \in L(R_+, R)$;
2. $k(t)$ is given by (3.1), with $\rho > 0$ and $k_0(t)$, $k_0'(t) \in L(R_+, R)$;
3. $\varphi(\sigma)$ is continuous from R into itself and satisfies

$$\sigma\varphi(\sigma) > 0 \qquad \text{for} \quad \sigma \neq 0; \tag{3.2}$$

4. there exists $q \geq 0$ such that

$$\text{Re}\{(1 - isq)G(s)\} \leq 0, \qquad s \neq 0, \tag{3.3}$$

where

$$G(s) = \tilde{k}_0(s) + \rho(is)^{-1}, \qquad s \neq 0. \tag{3.4}$$

Then there exists at least one solution $\sigma(t)$ of Eq. (1.1) such that

$$\sigma(t) \in C_0(R_+, R).$$

Moreover, any solution $\sigma(t) \in C_c(R_+, R)$ of Eq. (1.1) belongs to $C_0(R_+, R)$.

The proof of Theorem 3.1 will be accomplished in several steps:

A.   We shall first prove that under conditions (1)–(4), any continuous (on R_+) solution of Eq. (1.1) belongs to $C_0(R_+, R)$, i.e.,

$$\lim_{t \to \infty} \sigma(t) = 0. \tag{3.5}$$

Indeed, consider the following auxiliary functions obtained by the truncation procedure and convolution operation:

$$\varphi_t(\tau) = \begin{cases} \varphi(\sigma(\tau)), & 0 \leq \tau \leq t \\ 0, & \tau > t, \end{cases} \tag{3.6}$$

and

$$\lambda_t(\tau) = \int_0^\tau [k_0(\tau - u) + qk_0'(\tau - u)]\varphi_t(u) \, du + qk(0)\varphi_t(\tau), \tag{3.7}$$

each one being defined on R_+ for any fixed $t > 0$. Taking into account Eq. (1.1), the equation obtained from (1.1) by differentiation (almost everywhere on R_+) and (3.7), there results the following formulas for $\lambda_t(\tau)$:

$$\lambda_t(\tau) = \sigma(\tau) + q\sigma'(\tau) - [h(\tau) + qh'(\tau)] + \rho \int_0^\tau \varphi(\sigma(u)) \, du, \qquad 0 \leq \tau \leq t \tag{3.8}$$

and

$$\lambda_t(\tau) = \int_0^t [k_0(\tau - u) + qk_0'(\tau - u)]\varphi(\sigma(u)) \, du, \qquad \tau > t. \tag{3.9}$$

From (3.7) it follows that $\lambda_t(\tau) \in L \cap L^2$ on R_+, taking into account that the convolution product of a function from L by a function from L^p, $p \geq 1$, belongs to L^p. If we denote by $\tilde{\varphi}_t(s)$ and $\tilde{\lambda}_t(s)$ the Fourier transforms of $\varphi_t(s)$ and $\lambda_t(s)$, respectively, then we get

$$\tilde{\lambda}_t(s) = \int_0^\infty \lambda_t(\tau)e^{is\tau}\,d\tau = [(1 - isq)\tilde{k}_0(s) - q\rho]\tilde{\varphi}_t(s). \tag{3.10}$$

We now introduce a convenient function $\rho(t)$, given by

$$\rho(t) = \int_0^t \lambda_t(\tau)\varphi(\sigma(\tau))\,d\tau = \int_0^\infty \lambda_t(\tau)\varphi_t(\tau)\,d\tau. \tag{3.11}$$

By Parseval's formula we obtain

$$\rho(t) = (2\pi)^{-1} \int_R \tilde{\lambda}_t(s)\overline{\tilde{\varphi}_t(s)}\,ds. \tag{3.12}$$

Taking into account (3.10) and the fact that $\rho(t)$ is real, we can write (3.12) in the following form:

$$\rho(t) = (2\pi)^{-1} \int_R \text{Re}\{(1 - isq)\tilde{k}_0(s) - q\rho\}|\tilde{\varphi}_t(s)|^2\,ds. \tag{3.13}$$

But

$$\text{Re}\{(1 - isq)\tilde{k}_0(s) - q\rho\} = \text{Re}\{(1 - isq)G(s)\}, \tag{3.14}$$

with $G(s)$ defined by (3.4). Therefore, from (3.13), (3.14), and (3.3) we obtain $\rho(t) \leq 0$, $t > 0$. Starting now from (3.14) and taking into account (3.8), the inequality $\rho(t) \leq 0$ becomes

$$\int_0^t \varphi(\sigma(\tau))\sigma(\tau)\,d\tau + q\int_0^t \varphi(\sigma(\tau))\sigma'(\tau)\,d\tau$$

$$+ \rho\int_0^t\left\{\int_0^\tau \varphi(\sigma(u))\,du\right\}\varphi(\sigma(\tau))\,d\tau - \int_0^t[h(\tau) + qh'(\tau)]\varphi(\sigma(\tau))\,d\tau \leq 0. \tag{3.15}$$

If we denote

$$\phi(t) = \int_0^t \varphi(\sigma(\tau))\,d\tau, \qquad t \in R_+, \tag{3.16}$$

and

$$F(\sigma) = \int_0^\sigma \varphi(u)\,du, \qquad \sigma \in R, \tag{3.17}$$

then (3.15) yields

$$\int_0^t \varphi(\sigma(\tau))\sigma(\tau)\,d\tau + qF(\sigma(t)) + \tfrac{1}{2}\rho\phi^2(t)$$

$$- \int_0^t[h(\tau) + qh'(\tau)]\varphi(\sigma(\tau))\,d\tau - qF(\sigma(0)) \leq 0. \tag{3.18}$$

But

$$\int_0^t [h(\tau) + qh'(\tau)]\varphi(\sigma(\tau))\, d\tau = [h(t) + qh'(t)]\phi(t) - \int_0^t [h'(\tau) + qh''(\tau)]\phi(\tau)\, d\tau,$$

from which we get

$$\left| \int_0^t [h(\tau) + qh'(\tau)]\varphi(\sigma(\tau))\, d\tau \right| \le K \sup|\phi(\tau)|, \qquad 0 \le \tau \le t,$$

K being a positive number such that

$$\sup_{t \in R_+}\{|h(t)| + q|h'|\} + \int_0^\infty \{|h'(\tau)| + q|h''(\tau)|\}\, d\tau \le K.$$

Such a number K exists according to our condition (1). Observing that $\sigma(0) = h(0)$, we derive from (3.18) and the above estimates

$$\int_0^t \varphi(\sigma(\tau))\sigma(\tau)\, d\tau + qF(\sigma(t)) + \tfrac{1}{2}\rho\phi^2(t) - K \sup_{0 \le \tau \le t} |\phi(\tau)| - qF(h(0)) \le 0.$$

$$(3.19)$$

The first and second terms in (3.19) are nonnegative according to our assumption $\sigma\varphi(\sigma) > 0$ for $\sigma \ne 0$. Hence

$$\phi^2(t) - 2K\rho^{-1} \sup_{0 \le \tau \le t} |\phi(\tau)| - 2q\rho^{-1}F(h(0)) \le 0 \qquad (3.20)$$

for any $t > 0$. If we denote by $T(t)$ the largest real number such that

$$0 \le T(t) \le t, \ |\phi(T(t))| = \sup|\phi(\tau)|, \qquad 0 \le \tau \le t,$$

then replacing t by $T(t)$ in (3.20) we get

$$\phi(T(t)) \le K\rho^{-1} + [K^2\rho^{-2} + 2q\rho^{-1}F(h(0))]^{1/2}. \qquad (3.21)$$

Therefore, we can write

$$\phi(t) \le \alpha(F(h(0))), \qquad t \in R_+, \qquad (3.22)$$

which proves that $\phi(t)$ is bounded on R_+. We have put

$$\alpha(u) = K\rho^{-1} + [K^2\rho^{-2} + 2q\rho^{-1}u]^{1/2}, \qquad u \in R_+.$$

Consider now the following inequality derived from (3.19):

$$\int_0^t \varphi(\sigma(\tau))\sigma(\tau)\, d\tau \le K\alpha(F(h(0))) + qF(h(0)), \qquad t \in R_+. \qquad (3.23)$$

With the aim of applying Barbălat's lemma, we have to show that $\sigma(t)$ is uniformly continuous on R_+. Since Eq. (1.1) can be written in the form

$$\sigma(t) = h(t) + k(0)\phi(t) - \int_0^t k_0'(t - s)\phi(s)\, ds,$$

we obtain for $t \geq 0$

$$|\sigma(t)| \leq M + [|k(0)| + \int_0^\infty |k_0'(t)| \, dt]\alpha(F(h(0))) = \beta(F(h(0))), \quad (3.24)$$

with $M > 0$ such that $|h(t)| \leq M$ on R_+. Inequality (3.24) shows that $\sigma(t)$ is bounded on R_+. By differentiating both sides of (1.1) we get

$$\sigma'(t) = h'(t) + k(0)\varphi(\sigma(t)) + \int_0^t k_0'(t - s)\varphi(\sigma(s)) \, ds,$$

which proves that $\sigma'(t)$ is also bounded on R_+. Hence $\sigma(t)$ is uniformly continuous on R_+ and the same property holds for $\varphi(\sigma(t))\sigma(t)$. Therefore, the lemma of Barbălat is applicable to (3.23). There results $\varphi(\sigma(t))\sigma(t) \to 0$ as $t \to \infty$, from which we obtain (3.5). Assertion A is thus proved.

   B.   If $h(t)$ and $k(t)$ are continuous on R_+ and $\varphi(\sigma)$ satisfies the Lipschitz condition

$$|\varphi(\sigma) - \varphi(\xi)| \leq L|\sigma - \xi|, \qquad \sigma, \xi \in R, \quad (3.25)$$

then Eq. (1.1) has a unique continuous solution defined on R_+.

   The proof of statement B can be done by the classical method of successive approximations. The sequence

$$\sigma_0(t) = h(t), \qquad \sigma_{n+1}(t) = h(t) + \int_0^t k(t - s)\varphi(\sigma_n(s)) \, ds, \qquad n \geq 1,$$

consists of functions belonging to $C_c(R_+, R)$ and converges in the topology of this space (i.e., uniformly on any compact interval of R_+). The details of the proof are left to the reader.

   C.   Let $\varphi(\sigma)$ be a continuous mapping from R into itself such that (3.2) holds. If $T > 0$ is given, there exists a sequence $\{\varphi_n(\sigma)\}$ consisting of functions with the same properties as $\varphi(\sigma)$, $\varphi_n(\sigma) \to \varphi(\sigma)$ uniformly on $[-T, T]$ and such that each $\varphi_n(\sigma)$ satisfies a Lipschitz condition on R.

   Indeed, the functions $\varphi_n(\sigma)$ can be constructed as follows. If n is fixed, let us divide the interval $[-T, T]$ into $2n$ equal subintervals by the points $\sigma_k = kTn^{-1}$, $|k| = 0, 1, 2, \ldots, n$. $\varphi_n(\sigma)$ is the continuous function whose graph on $[-T, T]$ is the broken line with its successive vertices at $(\sigma_k, \varphi(\sigma_k))$, $\varphi_n(\sigma) = \varphi(-T)$ for $\sigma < -T$ and $\varphi_n(\sigma) = \varphi(T)$ for $\sigma > T$. It is obvious that $\varphi_n(\sigma)$ satisfies the required properties.

   D.   Under conditions (1)–(4), there exists at least one continuous (on R_+) solution of Eq. (1.1).

   Let $T > \beta(F(h(0)))$ be given, the function β being that defined by (3.24). Consider a sequence $\{\varphi_n(\sigma)\}$ as described in C, with T fixed as above. If we denote $F_n(\sigma) = \int_0^\sigma \varphi_n(u) \, du$, we shall have $T > \beta(F_n(h(0)))$ for sufficiently large

n. Without loss of generality we can assume that $T > \beta(F_n(h(0)))$ for any $n \geq 1$. The integral equation

$$\sigma(t) = h(t) + \int_0^t k(t - s)\varphi_n(\sigma(s)) \, ds \tag{3.26}$$

has a unique solution $\sigma_n(t)$, continuous on R_+. As seen above,

$$|\sigma_n(t)| \leq \beta(F_n(h(0))) < T.$$

If we differentiate both sides of (3.26), we easily find an estimate of the form $|\sigma_n'(t)| \leq T_1$, with T_1 independent of n. Therefore $\{\sigma_n(t)\}$ is uniformly bounded and equicontinuous on R_+. Then there exists a subsequence that converges uniformly on any compact interval of R_+. Let us assume that the sequence $\{\sigma_n(t)\}$ itself is uniformly convergent on any compact interval. From

$$\sigma_n(t) = h(t) + \int_0^t k(t - s)\varphi_n(\sigma_n(s)) \, ds,$$

we obtain for $n \to \infty$

$$\sigma(t) = h(t) + \int_0^t k(t - s)\varphi(\sigma(s)) \, ds,$$

i.e., Eq. (1.1) is satisfied by $\sigma(t) = \lim \sigma_n(t)$. Theorem 3.1 is thus proved.

We shall now apply the above result to the stability problem of the system of automatic control (1.8). The equivalent integral equation corresponds to $h(t)$ and $k(t)$ given by (1.9) and (1.10). It is obvious that conditions (1) and (2) from Theorem 1.1 are satisfied. For $\varphi(\sigma)$ we keep condition (3) from Theorem 1.1. Concerning condition (4), it is useful to remark that

$$\tilde{k}_0(s) = -(c, (isI + A)^{-1}b), \tag{3.27}$$

which says that the frequency criterion of absolute stability can be written as

$$\mathrm{Re}\{(isq - 1)[(c, (isI + A)^{-1}b) - \rho(is)^{-1}]\} \leq 0, \qquad s \neq 0. \tag{3.28}$$

As pointed out in Section 3.1, the system (1.8) describes both kinds of control: direct and indirect. Since A is a stable matrix, from the first equation of system (1.8) there results that $\lim_{t \to \infty} x(t) = 0$, for any initial conditions. From the last equation of (1.8), one sees that $\xi(t)$ also tends to zero as $t \to \infty$.

3.4 Further Investigation of the Preceding Case

We shall again consider Eq. (1.1), essentially under the same assumption regarding the kernel as in the preceding section. More precisely, we keep condition (3) from Theorem 3.1 and slightly modify conditions (1) and (2).

Condition (4) from Theorem 3.1 will be replaced—following Lellouche [1]—
by a more sophisticated inequality that involves two nonnegative parameters
q_1, q_2. It turns out that under the new frequency condition, the behavior of
the solutions is either that corresponding to asymptotic stability, or—and this
is the interesting feature—a kind of oscillation on $[0, \infty)$.

Theorem 4.1 Consider the equation

$$\sigma(t) = h(t) + \int_0^t k(t - s)\varphi(\sigma(s)) \, ds, \qquad t \in R_+, \tag{1.1}$$

and assume that the following conditions hold:

1. $h'(t)$, $h''(t) \in L(R_+, R)$ and $\int_0^t h(s) \, ds$ is bounded on R_+;
2. $k(t) = k_0(t) - \rho$, with $\rho > 0$, $k_0(t)$, $k_0'(t) \in L(R_+, R)$, and moreover,
$k_1(t) \in L(R_+, R)$, where

$$k_1(t) = - \int_t^\infty k_0(u) \, du; \tag{4.1}$$

3. $\varphi(\sigma)$ is a continuous function from R into itself such that

$$\sigma\varphi(\sigma) > 0, \qquad \sigma \neq 0; \tag{4.2}$$

4. there exist two real numbers q_1, $q_2 \geq 0$ such that

$$\mathrm{Re}\{[1 - isq_1 - (is)^{-1}q_2]\tilde{k}_0(s) - q_1\rho\} \leq 0 \tag{4.3}$$

for all real $s \neq 0$, the number q_2 also satisfying

$$\rho - q_2 \tilde{k}_0(0) > 0. \tag{4.4}$$

Then all solutions of Eq. (1.1) are defined on R_+ and have the following
property: either $\lim_{t \to \infty} \sigma(t) = 0$, or $\sigma(t) \pm \eta$, with $\eta > 0$ arbitrarily small, has
an infinite number of zeros on any half-axis $[T, \infty)$.

Proof Let $\sigma(t)$ be a solution of Eq. (1.1) and denote by $\varphi_t(\tau)$ the function
obtained by truncation from $\varphi(\sigma(\tau))$. It is obvious that $\varphi_t(\tau)$ is defined for all
$t \geq 0$ such that $\sigma(t)$ exists. Consider now the function $\lambda_t(\tau)$ given by

$$\lambda_t(\tau) = \int_0^\tau [k_0(\tau - u) + q_1 k_0'(\tau - u) + q_2 k_1(\tau - u)]\varphi_t(u) \, du$$

$$+ q_1 k(0)\varphi_t(\tau), \qquad \tau \in R_+, \tag{4.5}$$

where $k_1(t)$ is defined by (4.1). Since $\varphi_t(\tau) \in L \cap L^2$ on R_+, and k_0, k_0',
$k_1 \in L(R_+, R)$, there results that $\lambda_t(\tau) \in L \cap L^2$. Let us now find $\tilde{\lambda}_t(s)$. We have

$$\tilde{\lambda}_t(s) = \{\tilde{k}_0(s) + q_1[-is\tilde{k}_0(s) - k_0(0)]$$

$$+ q_2(is)^{-1}[-\tilde{k}_0(s) + \tilde{k}_0(0)]\}\tilde{\varphi}_t(s) + q_1 k(0)\tilde{\varphi}_t(s),$$

if we take into account that

$$\tilde{k}_1(s) = (is)^{-1}[-\tilde{k}_0(s) + \tilde{k}_0(0)].$$

It follows easily from the above formula for $\tilde{\lambda}_t(s)$ that

$$\tilde{\lambda}_t(s) = \{[1 - isq_1 - (is)^{-1}q_2]\tilde{k}_0(s) - q_1\rho + q_2(is)^{-1}\tilde{k}_0(0)\}\tilde{\varphi}_t(s). \quad (4.6)$$

Let us now consider the real function

$$\rho(t) = \int_0^t \lambda_t(\tau)\varphi(\sigma(\tau))\, d\tau = \int_0^\infty \lambda_t(\tau)\varphi_t(\tau)\, d\tau, \quad (4.7)$$

which by Parseval's relation can be written

$$\rho(t) = (2\pi)^{-1} \int_R \{[1 - isq_1 - (is)^{-1}q_2]\tilde{k}_0(s) - q_1\rho + q_2(is)^{-1}\tilde{k}_0(0)\}\,|\tilde{\varphi}_t(s)|^2\, ds.$$

Since $\rho(t)$ is real, we can obviously write

$$\rho(t) = (2\pi)^{-1} \int_R \operatorname{Re}\{[1 - isq_1 - (is)^{-1}q_2]\tilde{k}_0(s) - q_1\rho\}\,|\tilde{\varphi}_t(s)|^2\, ds. \quad (4.8)$$

According to our condition (4.2), we get from (4.8) that $\rho(t) \le 0$ for all $t \ge 0$ for which it is defined.

In order to give a convenient form to the condition $\rho(t) \le 0$, we shall replace some integrals in (4.5) by appropriate equivalent quantities. Namely, from (1.1) we have

$$\int_0^\tau k_0(\tau - u)\varphi(\sigma(u))\, du = \sigma(\tau) - h(\tau) + \rho\int_0^\tau \varphi(\sigma(u))\, du,$$

and from the equation obtained differentiating (1.1) we obtain

$$\int_0^\tau k_0'(\tau - u)\varphi(\sigma(u))\, du = \sigma'(\tau) - h'(\tau) - k(0)\varphi(\sigma(\tau)).$$

If we integrate both sides of (1.1) from 0 to $\tau > 0$, there results an equation which yields

$$\int_0^\tau k_1(\tau - u)\varphi(\sigma(u))\, du = \int_0^\tau \sigma(u)\, du - \int_0^\tau h(u)\, du$$

$$+ \rho\int_0^\tau (\tau - u)\varphi(\sigma(u))\, du - \tilde{k}_0(0)\int_0^\tau \varphi(\sigma(u))\, du.$$

Elementary calculations show that

$$\lambda_t(\tau) = \sigma(\tau) + q_1\sigma'(\tau) + q_2 \int_0^\tau \sigma(u)\, du$$

$$- \left[h(\tau) + q_1 h'(\tau) + q_2 \int_0^\tau h(u)\, du \right]$$

$$+ [\rho - q_2 \tilde{k}_0(0)] \int_0^\tau \varphi(\sigma(u))\, du + \rho \int_0^\tau (\tau - u)\varphi(\sigma(u))\, du \qquad (4.9)$$

for $0 \leq \tau \leq t$. From (4.7), (4.9), and $\rho(t) \leq 0$, we obtain the following (somewhat intricate but useful) basic inequality:

$$\int_0^\tau \varphi(\sigma(u))\sigma(u)\, du + q_1 F(\sigma(\tau)) + q_2 \int_0^\tau \varphi(\sigma(u))\, du \int_0^u \sigma(s)\, ds$$

$$+ q_2 \rho \int_0^\tau \varphi(\sigma(u))\, du \int_0^u (u - s)\varphi(\sigma(s))\, ds + \tfrac{1}{2}[\rho - q_2 \tilde{k}_0(0)]\phi^2(\tau)$$

$$- \int_0^\tau g(u)\varphi(\sigma(u))\, du - q_1 F(\sigma(0)) \leq 0, \qquad (4.10)$$

where $g(\tau) = h(\tau) + q_1 h'(\tau) + q_2 \int_0^\tau h(u)\, du$. We shall now derive some information concerning the solution $\sigma(t)$, starting from (4.10). Let us remark that the
first two integrals are nonnegative, according to our condition (3).

Assume that $\sigma(t)$ has only a fixed sign. This implies that $\varphi(\sigma(t))$ has a fixed
sign too and, consequently, the third and the fourth integrals in (4.10) are
positive. Therefore, from (4.10) we obtain

$$\tfrac{1}{2}[\rho - q_2\tilde{k}_0(0)]\phi^2(\tau) - \int_0^\tau g(u)\varphi(\sigma(u))\, du - q_1 F(\sigma(0)) \leq 0$$

for all τ such that $0 \leq \tau \leq t$. The preceding inequality leads to

$$\tfrac{1}{2}[\rho - q_2 \tilde{k}_0(0)]\phi^2(\tau) - K \sup_{0 \leq \tau \leq t} |\phi(\tau)| - q_1 F(\sigma(0)) \leq 0, \qquad (4.11)$$

where K is such that

$$\left| \int_0^t g(u)\varphi(\sigma(u))\, du \right| \leq K \sup_{0 \leq \tau \leq t} |\phi(\tau)|. \qquad (4.12)$$

In order to obtain (4.12), it suffices to integrate by parts in the integral appearing in the left side of (4.12) and to consider condition (1) of Theorem 4.1.
It is obvious that K does not depend on t. Inequality (4.11) has the same form
as (3.20) and, as seen in the preceding section, there results

$$|\phi(\tau)| \leq K_1, \qquad 0 \leq \tau \leq t. \qquad (4.13)$$

In other words, $\int_0^t \varphi(\sigma(u))\,du$ is bounded on the interval of existence of the solution $\sigma(t)$ considered above. By the same argument we encountered in the proof of Theorem 3.2, we obtain from (4.13) and Eq. (1.1) that $\sigma(t)$ is bounded on its interval of existence and is uniformly continuous (more precisely, there results also the boundedness of the derivative $\sigma'(t)$). This implies, first, that a finite escape time for $\sigma(t)$ cannot exist. Secondly, Barbălat's lemma applies, and we find that $\sigma(t)$ tends asymptotically to zero.

Suppose now that $\sigma(t)$ has sign changes only for $t \in [0, t_1]$. Therefore, for $t > t_1 > 0$, with t_1 conveniently chosen, $\sigma(t)$ has a fixed sign. There are two distinct situations to be discussed. The first one corresponds to the case when $\phi(t)$ is bounded on the interval of existence for $\sigma(t)$. Then $\int_{t_1}^t \varphi(\sigma(u))\,du$ is bounded too, which leads again to the conclusion that $\sigma(t)$ goes to zero as $t \to \infty$. If $\phi(t)$ is unbounded, then the last three terms in (4.10) sum to a positive value for $t > t^* > t_1$. Furthermore, if we assume that the statement is not true, there exists $\eta > 0$, such that the third integral in the left-hand side of (4.10) can be written as

$$I_3 = \int_{t_1+\varepsilon}^t \varphi(\sigma(u))\,du \int_{t_1+\varepsilon}^u \sigma(s)\,ds + \int_0^{t_1+\varepsilon} \sigma(s)\,ds \int_{t_1+\varepsilon}^t \varphi(\sigma(u))\,du$$

$$+ \int_0^{t_1+\varepsilon} \varphi(\sigma(u))\,du \int_0^u \sigma(s)\,ds,$$

where $\varepsilon > 0$ is chosen such that $|\sigma(t)| > \eta$ and $|\varphi(\sigma(t))| > \delta$ for $t > t_1 + \varepsilon$. Assume for instance that $\sigma(t) > 0$ for $t > t_1$. Then we get the following lower estimate for $t > t_1 + \varepsilon$:

$$I_3 > \tfrac{1}{2}(t - t_1 - \varepsilon)^2 \eta\delta + \delta(t - t_1 - \varepsilon)\int_0^{t_1+\varepsilon} \sigma(s)\,ds + \int_0^{t_1+\varepsilon} \varphi(\sigma(u))\,du \int_0^u \sigma(s)\,ds.$$

For sufficiently large t, I_3 becomes and remains thereafter positive (similarly, if $\sigma(t) < 0$, for $t > t_1$). An equivalent analysis can be done for the fourth integral in the first member of (4.10). Therefore, for sufficiently large t we violate the inequality (4.10), the left-hand side being strictly positive. Hence, $\sigma(t)$ cannot behave otherwise than described in Theorem 4.1.

Suppose, however, that $\sigma(t)$ has a finite escape time at $\tilde{t} > t_1$, after a finite number of zeros, t_1 being the last one. From Eq. (1.1) there results

$$|\sigma(t)| \le K_2 + K_3 \int_0^t |\varphi(\sigma(s))|\,ds$$

for appropriate K_2, $K_3 > 0$, as long as $\sigma(t)$ is finite. The above inequality implies that $\int_{t_1}^{\tilde{t}} \varphi(\sigma(u))\,du$ diverges, i.e., $\phi(t) \to \infty$ as $t \to \tilde{t}$. We shall now examine the sign of the third and fourth integrals in (4.10). We have

$$I_3 = \int_{t_1}^t \varphi(\sigma(u)) \, du \int_{t_1}^u \sigma(s) \, ds + \int_0^{t_1} \sigma(s) \, ds \int_{t_1}^t \varphi(\sigma(u)) \, du$$

$$+ \int_0^{t_1} \varphi(\sigma(u)) \, du \int_0^u \sigma(s) \, ds.$$

If we take the sign of $\sigma(t)$ (and hence of $\varphi(\sigma(t))$) to be positive for $\tilde{t} > t > t_1$, then the first term on the right is positive. Now we add the first integral in (4.10) to the third and find

$$I = \int_0^t \varphi(\sigma(u))\sigma(u) \, du + q_2 \int_0^t \varphi(\sigma(u)) \, du \int_0^u \sigma(s) \, ds$$

$$> \int_{t_1}^t \varphi(\sigma(u)) \, du \left[\sigma(u) + q_2 \int_0^u \sigma(s) \, ds \right] + q_2 \int_0^{t_1} \varphi(\sigma(u)) \, du \int_0^u \sigma(s) \, ds.$$

The integral $\int_0^{t_1} \sigma(s) \, ds$ is a constant, while $\sigma(u)$ is of fixed sign for $t_1 < u < \tilde{t}$ and diverges for $u \to \tilde{t}$. Hence $\sigma(u) + q_2 \int_0^u \sigma(s) \, ds > \eta > 0$ for $u \geq t_2 > t_1$, assuming that $\sigma(u)$ is positive for $u > t_1$. Therefore

$$I > \eta \int_{t_2}^t \varphi(\sigma(u)) \, du + \int_{t_1}^{t_2} \varphi(\sigma(u)) \left[\sigma(u) + q_2 \int_0^u \sigma(s) \, ds \right] du$$

$$+ q_2 \int_0^{t_1} \varphi(\sigma(u)) \, du \int_0^u \sigma(s) \, ds,$$

from which we get that the sum of the first and third integrals in (4.10) tends to infinity as $t \to \tilde{t}$. We can handle the fourth integral in (4.10) as follows:

$$I_4 = \int_0^t \varphi(\sigma(u)) \, du \int_0^u (u - s)\varphi(\sigma(s)) \, ds$$

$$= \int_0^{t_1} \varphi(\sigma(u)) \, du \int_0^u (u - s)\varphi(\sigma(s)) \, ds + \int_{t_1}^t \varphi(\sigma(u)) \, du \int_0^{t_1} (u - s)\varphi(\sigma(s)) \, ds$$

$$+ \int_{t_1}^t \varphi(\sigma(u)) \, du \int_{t_1}^u (u - s)\varphi(\sigma(s)) \, ds.$$

The third term is positive since $\varphi(\sigma(u))$ has a fixed sign for $u > t_1$, while the first term is a constant. We add the second term from I_4, multiplied by $q_2 \rho$, to I and find easily that

$$I + \int_{t_1}^t \varphi(\sigma(u)) \, du \int_0^{t_1} (u - s)\varphi(\sigma(s)) \, ds$$

tends to ∞ as $t \to \tilde{t}$. Since the second term in the left side of (4.10) is always positive, there results that (4.10) will again be violated for t sufficiently near $\tilde{t}$. Of course, we have to consider that the last three terms in (4.10) sum up to a positive value for t sufficiently close to $\tilde{t}$, because $\phi(t) \to \infty$ as $t \to \tilde{t}$. Therefore, a finite escape time cannot exist.

Theorem 4.1 is now completely proved, since $\sigma(t)$ must either go to zero asymptotically, or else $\sigma(t) \pm \eta$, with arbitrarily small $\eta > 0$, has infinitely many zeros on any half-axis $[T, \infty)$.

Remark   As shown in Theorem 3.2, if we take $q_2 = 0$ in our theorem, then the solution goes asymptotically to zero. Therefore, oscillatory solutions could appear only if (4.3) and (4.4) are verified for some $q_2 > 0$.

A problem that remains open is to find supplementary conditions under which the existence of oscillatory solutions is guaranteed.

For $\varphi(\sigma) = a\sigma$, some considerations can be found in the paper [1] by Lellouche. They show that the boundedness of the solution does not follow necessarily from the conditions of Theorem 4.1.

3.5   Another Case of a Nonintegrable Kernel

In this section we shall again be concerned with Eq. (1.1), under the following basic condition on the kernel:

$$k(t) = k_0(t) + \alpha \cos \omega t + \beta \sin \omega t, \tag{5.1}$$

with $k_0(t) \in L(R_+, R)$. Such kernels are encountered in the integral equation obtained for $\sigma(t)$ in the case of a direct control system of the form (1.4), if the matrix A has a pair of conjugate pure imaginary characteristic roots and the remaining roots have negative real parts. In other words, a critical case of stability has to be discussed with respect to the system (1.4).

The main result of this section can be stated as follows.

Theorem 5.1   Consider equation

$$\sigma(t) = h(t) + \int_0^t k(t - s)\varphi(\sigma(s)) \, ds, \qquad t \in R_+, \tag{1.1}$$

under the following assumptions:

1. $h(t)$ can be represented in the form

$$h(t) = h_0(t) + \lambda \cos \omega t + \mu \sin \omega t, \qquad \omega \neq 0, \tag{5.2}$$

with $h_0(t), h_0'(t) \in L(R_+, R)$;

2. $k(t)$ is of the form (5.1) with $k_0(t), k_0'(t) \in L(R_+, R)$, $\omega \neq 0$, $\alpha < 0$, $\beta \leq 0$;

3. $\varphi(\sigma)$ is a continuous and bounded mapping from R into itself such that

$$0 < \sigma\varphi(\sigma) < \gamma\sigma^2, \qquad \sigma \neq 0, \tag{5.3}$$

with $0 < \gamma \leq \infty$;

4. the frequency condition

$$-\gamma^{-1} + (\beta/\omega) + \text{Re}\{[1 - (is\beta/\omega\alpha)]\tilde{k}_0(s)\} \le 0 \qquad (5.4)$$

holds for all $s \in R$.

Then $\sigma(t)$ tends to zero as $t \to \infty$.

Proof Equation (1.1) can be written as

$$\sigma(t) = h_0(t) + \int_0^t k_0(t - s)\varphi(\sigma(s))\, ds + u(t), \qquad (5.5)$$

with

$$u(t) = \lambda \cos \omega t + \mu \sin \omega t + \int_0^t \vartheta(t - s)\varphi(\sigma(s))\, ds, \qquad (5.6)$$

where

$$\vartheta(t) = \alpha \cos \omega t + \beta \sin \omega t. \qquad (5.7)$$

Let θ be an arbitrary fixed number. We shall prove that $u(t)$ can be represented in the form

$$u(t) = \eta \cos \theta + \zeta \sin \theta, \qquad (5.8)$$

with $\eta(t)$ and $\zeta(t)$ uniquely determined by

$$\begin{aligned} \eta' &= \omega\zeta + \alpha_1\varphi(\sigma), & \eta(0) &= \eta_0, \\ \zeta' &= -\omega\eta + \alpha_2\varphi(\sigma), & \zeta(0) &= \zeta_0, \end{aligned} \qquad (5.9)$$

where

$$\begin{aligned} \alpha_1 &= \alpha \cos \theta - \beta \sin \theta, & \alpha_2 &= \alpha \sin \theta + \beta \cos \theta, \\ \eta_0 &= \lambda \cos \theta - \mu \sin \theta, & \zeta_0 &= \lambda \sin \theta + \mu \cos \theta. \end{aligned} \qquad (5.10)$$

Indeed, from (5.9) we obtain

$$\eta(t) = \eta_0 \cos \omega t + \zeta_0 \sin \omega t + \int_0^t [\alpha_1 \cos \omega(t - s) + \alpha_2 \sin \omega(t - s)]\varphi\, ds,$$

$$\zeta(t) = \zeta_0 \cos \omega t - \eta_0 \sin \omega t + \int_0^t [\alpha_2 \cos \omega(t - s) - \alpha_1 \sin \omega(t - s)]\varphi\, ds.$$

From (5.6), (5.7), (5.8), and the above formulas for $\eta(t)$ and $\zeta(t)$, (5.10) follows easily.

Consider now the following auxiliary function

$$\rho(t) = \int_0^\infty [\sigma(\tau)\varphi_t(\tau) - \gamma^{-1}\varphi_t^2(\tau)]\, d\tau, \qquad (5.11)$$

where, as usual, $\varphi_t(\tau)$ is the function obtained by the truncation of $\varphi(\sigma(\tau))$. If we denote

$$\sigma_1(\tau) = \int_0^\tau k_0(\tau - u)\varphi_t(u)\, du, \tag{5.12}$$

then (5.5) yields for $0 \le \tau \le t$

$$\sigma(\tau) = h_0(\tau) + \sigma_1(\tau) + u(\tau), \tag{5.13}$$

and (5.11) becomes

$$\rho(t) = \int_0^\infty h_0(\tau)\varphi_t(\tau)\, d\tau + \int_0^\infty [\sigma_1(\tau)\varphi_t(\tau) - \gamma^{-1}\varphi_t^{\,2}(\tau)]\, d\tau + \int_0^\infty u(\tau)\varphi_t(\tau)\, d\tau. \tag{5.14}$$

But from (5.10) one derives easily

$$u\varphi_t = (\eta \cos \theta + \zeta \sin \theta)\varphi_t = \alpha^{-1}(\alpha_1\eta + \alpha_2\zeta)\varphi_t - (\beta/\alpha)(\zeta \cos \theta - \eta \sin \theta)\varphi_t. \tag{5.15}$$

If $0 \le \tau \le t$, from (5.9) there results

$$(\alpha_1\eta + \alpha_2\zeta)\varphi_t = \tfrac{1}{2}(d/d\tau)(\eta^2 + \zeta^2). \tag{5.16}$$

By differentiation of (5.13) we obtain

$$\sigma'(\tau) = h_0'(\tau) + \sigma_1'(\tau) + (\eta' \cos \theta + \zeta' \sin \theta),$$

and taking into account (5.9) we get

$$\sigma'(\tau) = h_0'(\tau) + \sigma_1'(\tau) + \omega(\zeta \cos \theta - \eta \sin \theta) + \alpha\varphi_t(\tau), \tag{5.17}$$

for $0 \le \tau \le t$. From (5.14)–(5.17), we obtain

$$\rho(t) = \int_0^t [h_0(\tau) + (\beta/\omega\alpha)h_0'(\tau)]\varphi(\sigma(\tau))\, d\tau + (2\alpha)^{-1}(\eta^2 + \zeta^2)\Big|_0^t$$

$$- (\omega\alpha)^{-1} \int_0^t \varphi(\sigma(\tau))\sigma'(\tau)\, d\tau + I, \tag{5.18}$$

where

$$I = \int_0^\infty \{[\sigma_1(\tau) + (\beta/\omega\alpha)\sigma_1'(\tau)]\varphi_t(\tau) + (-\gamma^{-1} + (\beta/\omega)\varphi_t^{\,2}(\tau)\}\, d\tau.$$

By Parseval's equality, it follows that I can be also written in the form

$$I = (2\pi)^{-1} \int_R \{-\gamma^{-1} + (\beta/\omega) + \mathrm{Re}[(1 - (is\beta/\omega\alpha))\tilde{k}_0(s)]\}|\tilde{\varphi}_t(s)|^2\, ds. \tag{5.19}$$

Let us remark that both σ_1 and σ_1' belong to $L \cap L^2$ on R_+, by our hypotheses.

According to (5.4), we have $I \le 0$, which yields

$$\rho(t) - (2\alpha)^{-1}[\eta^2(t) + \zeta^2(t)] + (\beta/\omega\alpha)F(\sigma(t)) \le -(2\alpha)^{-1}(\eta_0^2 + \zeta_0^2)$$

$$+ (\beta/\omega\alpha)F(\sigma(0)) + \int_0^t [h_0(\tau) + (\beta/\omega\alpha)h'(\tau)]\varphi(\dot{\sigma}(\tau))\, d\tau$$

if we observe (5.18). As usual, $F(\sigma) = \int_0^\sigma \varphi(u)\, du$. Replacing $\rho(t)$ in the above inequality by its equivalent form (5.11), we get

$$\int_0^t \varphi(\sigma(\tau))[\sigma(\tau) - \gamma^{-1}\varphi(\sigma(\tau))]\, d\tau - (2\alpha)^{-1}[\eta^2(t) + \zeta^2(t)] + (\beta/\omega\alpha)F(\sigma(t)$$

$$\le -(2\alpha)^{-1}(\eta_0^2 + \zeta_0^2) + (\beta/\omega\alpha)F(\sigma(0))$$

$$+ \int_0^t [h_0(\tau) + (\beta/\omega\alpha)h_0'(\tau)]\varphi(\sigma(\tau))\, d\tau. \tag{5.20}$$

This is the basic inequality in studying the behavior of $\sigma(t)$ at infinity. All the quantities in the left-hand side are nonnegative. Namely, the first term has a nonnegative integrand, according to (5.3). Indeed, it can be written as $(\varphi/\sigma)[\sigma^2 - \gamma^{-1}\sigma\varphi]$ for $\sigma \neq 0$, and both factors are positive. The integrand is zero for $\sigma = 0$. The second and third terms in the left-hand side of (5.20) are nonnegative because $\alpha < 0$, $\beta \le 0$, and $F(\sigma) > 0$ for $\sigma \neq 0$. It remains only to handle the third term in the right-hand side of (5.20). From condition (1) and (3) there results that

$$\left| \int_0^t [h_0(\tau) + (\beta/\omega\alpha)h_0'(\tau)]\varphi(\sigma(\tau))\, d\tau \right| \le \phi \int_0^\infty (|h_0(\tau)| + (\beta/\omega\alpha)|h_0'(\tau)|)\, d\tau,$$

where ϕ is such that $|\varphi(\sigma)| \le \phi$, $\sigma \in R$. Therefore, the right-hand side of inequality (5.20) is bounded on R_+. In particular, we have for a convenient $K > 0$

$$\int_0^t [\varphi(\sigma(\tau)) - \gamma^{-1}\varphi^2(\sigma(\tau))]\sigma(\tau)\, d\tau \le K, \qquad t \in R_+. \tag{5.21}$$

In order to apply Barbălat's lemma, we shall prove that $\sigma(t)$ is uniformly continuous on R_+. This property will follow from the fact that the derivative $\sigma'(t)$, which is given by (5.17), is dominated by the sum of an integrable function on R_+ and a bounded one [the first is $|h_0'(\tau) + \sigma_1'(\tau)|$, and the second is $\omega(\zeta \cos\theta - \eta \sin\theta) + \alpha\varphi$]. That η and ζ are bounded on R_+ one can see from the same inequality (5.20). Hence, from (5.21) we obtain the desired stability result: $\lim_{t\to\infty} \sigma(t) = 0$.

Remark Further arguments show that $\lim_{t\to\infty} \eta(t) = 0$, $\lim_{t\to\infty} \zeta(t) = 0$. In other words, the system consisting of Eq. (5.5), (5.8), and (5.9) is asymptotically stable.

3.6   A Stability Result in the Vector Case

The idea of considering vector equations of the form (1.1) is related to the necessity of studying automatic control systems with several nonlinearities (see, for instance Lefschetz [1]).

Consider again equation

$$\sigma(t) = h(t) + \int_0^t k(t - s)\varphi(\sigma(s)) \, ds, \qquad t \in R_+, \tag{1.1}$$

with $\sigma(t)$ a vector function ($\sigma \in R^n$). Then $h(t)$ should also be an n-vector function, while $k(t)$ is a matrix kernel of type n by n. Concerning $\varphi(\sigma)$, we will make convenient hypotheses below. The result we shall establish is due to Doležal [1]. It constitutes a direct extension of the Popov's approach to the case of vector-valued nonlinearities.

Theorem 6.1   Consider Eq. (1.1) under the following assumptions:

1. $h'(t)$, $h''(t) \in L(R_+, R^n)$;
2. $k(t) = k_0(t) - \gamma$, with $k_0(t)$ an n by n matrix such that

$$\|k_0(t)\|, \ \|k_0{}'(t)\| \in L(R_+, R)$$

and γ a symmetric positive definite matrix of the same type;

3. there exists a real scalar function $U(\sigma)$, possessing everywhere on R^n continuous first-order partial derivatives such that $\varphi(\sigma) = \text{grad } U(\sigma)$, i.e.,

$$\varphi_i(\sigma) = (\partial U(\sigma)/\partial \sigma_i), \qquad i = 1, 2, \ldots, n; \tag{6.1}$$

4. there are two positive numbers α and β such that

$$\alpha\|\sigma\|^2 \le (\varphi(\sigma), \sigma), \qquad \|\varphi(\sigma)\| \le \beta\|\sigma\| \tag{6.2}$$

for any $\sigma \in R^n$;

5. if $c > \alpha^{-1}\beta^2$, there is a $q \ge 0$ such that the matrix

$$A(s) = (1 - isq)\tilde{k}_0(s) - (c^{-1}I + q\gamma) \tag{6.3}$$

satisfies the (frequency) condition

$$\text{Re}(\bar{\eta}, A(s)\eta) \le 0 \tag{6.4}$$

for any $s \in R$ and any complex n-vector η.

Then any solution of Eq. (1.1) tends to zero as $t \to \infty$.

Proof   Let $\sigma(t)$ be a continuous (on R_+) solution of (1.1). We are not concerned with the existence problem here. Before applying the truncation procedure and Parseval's inequality, it is useful to remark that (6.2) implies

$$\alpha\|\sigma\| \le \|\varphi(\sigma)\| \le \beta\|\sigma\|, \tag{6.5}$$

$$\alpha\|\sigma\|^2 \le (\varphi(\sigma), \sigma) \le \beta\|\sigma\|^2, \tag{6.6}$$

and also

$$(\varphi(\sigma), \sigma) - c^{-1}\|\varphi(\sigma)\|^2 \ge c_1\|\sigma\|^2, \tag{6.7}$$

with $c_1 = c^{-1}(\alpha c - \beta^2)$. Next, let us remark that the curvilinear integral of $(\varphi(\sigma), d\sigma)$ is independent of the path and we have

$$\int_{\sigma_1}^{\sigma_2} (\varphi(\sigma), d\sigma) = U(\sigma_2) - U(\sigma_1). \tag{6.8}$$

If for a given $\sigma \in R^n$ we take as integration path from the origin O to σ the line segment joining them, then we easily get

$$\tfrac{1}{2}\alpha\|\sigma\|^2 \le U(\sigma) - U(0) \le \tfrac{1}{2}\beta\|\sigma\|^2. \tag{6.9}$$

If $\varphi_t(\tau)$ denotes the function obtained by truncation from $\varphi(\sigma(\tau))$ in the usual way, let us consider the auxiliary function

$$\lambda_t(\tau) = \int_0^\infty k_0(\tau - u)\varphi_t(u)\, du. \tag{6.10}$$

Under our assumptions we have $\lambda_t, \lambda_t' \in L \cap L^2$ on R_+. Next, let us define

$$\rho(t) = \int_0^t (\varphi_t(\tau), \lambda_t(\tau) - c^{-1}\varphi_t(\tau) + q[\lambda_t'(\tau) - \gamma\varphi_t(\tau)])\, d\tau. \tag{6.11}$$

Of course, the integral in (6.11) can be written with the limits 0 and ∞. To (6.11) we can apply Parseval's equality and find

$$\rho(t) = (2\pi)^{-1} \int_R (\overline{\tilde{\varphi}_t(s)}, \tilde{\lambda}_t(s) - c^{-1}\tilde{\varphi}_t(s) + q[-is\tilde{\lambda}_t(s) - \gamma\tilde{\varphi}_t(s)])\, ds.$$

Taking into account that $\tilde{\lambda}_t(s) = \tilde{k}_0(s)\tilde{\varphi}_t(s)$, we obtain for $\rho(t)$ the following equivalent form:

$$\rho(t) = (2\pi)^{-1} \int_R (\overline{\tilde{\varphi}_t(s)}, [(1 - isq)\tilde{k}_0(s) - (c^{-1}I + q\gamma)]\tilde{\varphi}_t(s))\, ds.$$

From (6.3) and the fact $\rho(t)$ is real, there results

$$\rho(t) = (2\pi)^{-1} \int_R \mathrm{Re}(\overline{\tilde{\varphi}_t(s)}, A(s)\tilde{\varphi}_t(s))\, ds, \tag{6.12}$$

which—according to (6.4)—leads to $\rho(t) \le 0$ for all $t > 0$. But Eq. (1.1) gives

$$\int_0^\tau k_0(\tau - u)\varphi(\sigma(u))\, du = \sigma(\tau) - h(\tau) + \gamma \int_0^\tau \varphi(\sigma(u))\, du,$$

from which we get for $0 \leq \tau \leq t$

$$\lambda_t(\tau) = \sigma(\tau) - h(\tau) + \gamma \int_0^\tau \varphi(\sigma(u)) \, du. \tag{6.13}$$

Now, from (6.11), (6.13), and $\rho(t) \leq 0$ we obtain the following inequality:

$$\int_0^t (\varphi(\sigma(\tau)), \sigma(\tau) - c^{-1}\varphi(\sigma(\tau))) \, d\tau + q \int_0^t (\varphi(\sigma(\tau)), \sigma'(\tau)) \, d\tau$$

$$+ \int_0^t (\varphi(\sigma(\tau)), \gamma \int_0^\tau \varphi(\sigma(u)) \, du) \, d\tau$$

$$\leq \int_0^t (\varphi(\sigma(\tau)), h(\tau) + qh'(\tau)) \, d\tau.$$

If we denote as usual $\phi(t) = \int_0^t \varphi(\sigma(\tau)) \, d\tau$, we get further

$$\int_0^t (\varphi(\sigma(\tau)), \sigma(\tau) - c^{-1}\varphi(\sigma(\tau))) \, d\tau + q \int_{\sigma(0)}^{\sigma(t)} (\varphi(\sigma), d\sigma) + \tfrac{1}{2}(\phi(\tau), \gamma\phi(\tau)) \Big|_0^t$$

$$\leq (\phi(\tau), h(\tau) + qh'(\tau)) \Big|_0^t - \int_0^t (\phi(\tau), h'(\tau) + qh''(\tau)) d\tau.$$

If we consider (6.7), (6.8), (6.9), and condition (1) from the statement of our theorem, then the above inequality gives

$$c_1 \int_0^t \|\sigma(\tau)\|^2 \, d\tau + \tfrac{1}{2}q\alpha\|\sigma(t)\|^2 + \tfrac{1}{2}(\phi(t), \gamma\phi(t)) \leq \tfrac{1}{2}q\beta\|\sigma(0)\|^2 + M \sup_{0 \leq \tau \leq t} \|\phi(\tau)\|,$$

with $M > 0$ such that

$$\|h(t)\| + q\|h'(t)\| + \int_0^\infty [\|h'(\tau)\| + q\|h''(\tau)\|] \, d\tau \leq M$$

for $t \in R_+$. Such a M exists, according to condition (1). Since γ is a positive definite symmetric matrix, there exists $\mu > 0$ such that $(\eta, \gamma\eta) \geq \mu\|\eta\|^2$, for any n-vector η. Therefore, we obtain the following basic inequality:

$$c_1 \int_0^t \|\sigma(\tau)\|^2 \, d\tau + \tfrac{1}{2}q\alpha\|\sigma(t)\|^2 + \tfrac{1}{2}\mu\|\phi(t)\|^2 \leq \tfrac{1}{2}q\beta\|\sigma(0)\|^2 + M \sup_{0 \leq \tau \leq t} \|\phi(\tau)\|. \tag{6.14}$$

Since any term in the left-hand side of (6.14) is positive, we obtain as a special case

$$\tfrac{1}{2}\mu\|\phi(t)\|^2 \leq \tfrac{1}{2}q\beta\|\sigma(0)\|^2 + M \sup_{0 \leq \tau \leq t} \|\phi(\tau)\|,$$

which can be handled in the same manner as the inequality (3.20) in the proof of Theorem 3.1. It follows that $\phi(t)$ is bounded on R_+. This implies that the right member in (6.14) is bounded on R_+ and, consequently, any term in the first member of (6.14) is bounded on R_+. There results that $\sigma(t)$ is bounded and $\sigma \in L^2(R_+, R^n)$. In other words, $\sigma \in C \cap L^2$ on R_+. It remains to show that $\sigma \in C_0$. This will follow easily if we observe that

$$\sigma'(t) = h'(t) + \int_0^t k_0'(t - \tau)\varphi(\sigma(\tau))\,d\tau + k(0)\varphi(\sigma(t))$$

is bounded on R_+. Hence, $\sigma(t)$ is uniformly continuous on R_+ and, since $\sigma \in L^2$, we obtain that $\sigma \in C_0$ on R_+. Theorem 6.1 is thus proved.

Remark A very close result to that given in Theorem 6.1 and—at the same time—a straightforward generalization of Theorem 3.1 can be obtained if we slightly modify conditions (4) and (5) from Theorem 6.1. Namely, instead of (4) we can assume $(\varphi(\sigma), \sigma) > 0$ for any $\sigma \neq 0$. Accordingly, the matrix $A(s)$ used in the proof should be defined by

$$A(s) = (1 - isq)\tilde{k}_0(s) - q\gamma.$$

We shall now apply the above result to the system of automatic control described by

$$\dot{x} = Ax + B\varphi(\sigma), \qquad \dot{\zeta} = \varphi(\sigma), \qquad \sigma = Cx - \gamma\zeta.$$

This time, B, C, and γ are square matrices of type n by n, the last one being symmetric and positive. The nonlinearity is vector-valued: $\varphi, \zeta \in R^n$. From the first and second equations we obtain

$$x(t) = e^{At}x^0 + \int_0^t e^{A(t-s)}B\varphi(\sigma(s))\,ds,$$

$$\zeta(t) = \zeta^0 + \int_0^t \varphi(\sigma(s))\,ds.$$

Using the last equation of the system we get an integral equation of the form (1.1), with

$$h(t) = Ce^{At}x^0 - \gamma\zeta^0, \qquad k(t) = Ce^{At}B - \gamma.$$

Consequently, $k_0(t) = Ce^{At}B$. This leads to the following formula for $\tilde{k}_0(s)$:

$$\tilde{k}_0(s) = -C(isI + A)^{-1}B,$$

taking into account (2.20). Hence, the matrix $A(s)$ to be considered in the frequency condition corresponding to (6.4) is now

$$A(s) = (isq - 1)C(isI + A)^{-1}B - (c^{-1}I + q\gamma),$$

where γ and c have the same meaning as in the proof of Theorem 6.1.

3.7 An Integro-Differential Equation

In the theory of electric networks, the following differential system is encountered:

$$\dot{x} = -Ax + By, \qquad \dot{y} = Cx - f(y)$$

where x is an n-vector, y an m-vector, A, B, C are constant matrices, and $f(y)$ is an m-vector function. It is very easy to obtain for y an integro-differential equation of the form

$$\dot{y}(t) + \int_0^t k(t - s)\dot{y}(s)\, ds = -\varphi(y) + \lambda(t), \tag{7.1}$$

with

$$k(t) = CA^{-1}e^{-At}B,$$

$$\varphi(y) = f(y) - CA^{-1}By,$$

$$\lambda(t) = Ce^{-At}x^0, \qquad x^0 \in R.$$

Following Moser [1], we shall investigate the behavior of solutions of the integro-differential equation (7.1), under more general assumptions for $k(t)$, $\varphi(y)$, and $\lambda(t)$ than those listed above. The method we shall use leans upon Popov's frequency method. The basic condition is similar to that given above for vector integral equations. The result we shall prove in this section can be stated as follows.

Theorem 7.1 Consider the vector integro-differential equation (7.1) and assume that the following conditions hold:

1. $\varphi(y) = \text{grad } U(y)$, $y \in R^n$, the scalar function $U(y)$ being such that

$$\lim U(y) = 0 \qquad \text{as} \quad \|y\| \to \infty; \tag{7.2}$$

2. the matrix kernel $k(t)$ is such that $\|k(t)\| \in L^2(R_+, R)$;
3. $\lambda(t) \in C_0 \cap L^2$ on R_+, taking its values in R^n;
4. the frequency condition

$$\text{Re}\{\|\eta\|^2 + (\bar{\eta}, \tilde{k}(s)\eta)\} \geq \delta\|\eta\|^2, \qquad s \in R, \tag{7.3}$$

holds for some $\delta > 0$ and any complex n-vector η.

Then any solution $y(t)$ of Eq. (7.1) exists on R_+, and its limit set agrees with that of a solution of the differential system

$$\dot{z} = -\varphi(z). \tag{7.4}$$

Remark   Before proceeding to the proof of Theorem 7.1, we shall point out that $\tilde{k}(s)$ denotes the generalized Fourier transform of $k(t)$, as defined in Exercise 4, Section 1.3. Therefore, $\tilde{k}(s)$ is defined only almost everywhere on R.

Proof   Let $y(t)$ be a solution of (7.1) satisfying $y(0) = y^0 \in R^n$. We shall prove first that

$$\|y(t)\| \le c, \qquad \int_0^\infty \|\dot{y}(t)\|^2 \, dt \le c_1, \qquad (7.5)$$

where c and c_1 depend only upon y^0. Since

$$\int_0^t (\varphi(y(\tau)), \dot{y}(\tau)) \, d\tau = U(y(t)) - U(y^0),$$

we easily obtain from (7.1) that

$$\int_0^t (\dot{y} + K\dot{y}, \dot{y}) \, d\tau + U(y(t)) = \int_0^t (\lambda(\tau), \dot{y}) \, d\tau + U(y^0)$$

for all $t \ge 0$ such that $y(t)$ exists. In the above formula, K stands for the operator generated by the kernel $k(t)$. Parseval's equality and (7.3) yield

$$\int_0^t (\dot{y} + K\dot{y}, \dot{y}) \, d\tau \ge \delta \int_0^t \|\dot{y}(\tau)\|^2 \, d\tau$$

and thus

$$\delta \int_0^t \|\dot{y}(\tau)\|^2 \, d\tau + U(y(t)) \le \int_0^t (\lambda, \dot{y}) \, d\tau + U(y^0).$$

Estimating the integral on the right in a standard fashion, we get

$$\int_0^t (\lambda, \dot{y}) \, d\tau \le (\delta/2) \int_0^t \|\dot{y}(\tau)\|^2 \, d\tau + (2\delta)^{-1} \int_0^t \|\lambda(\tau)\|^2 \, d\tau.$$

Therefore, we have the following basic inequality:

$$(\delta/2) \int_0^t \|\dot{y}(\tau)\|^2 \, d\tau + U(y(t)) \le (2\delta)^{-1} \int_0^t \|\lambda(\tau)\|^2 \, d\tau + U(y^0). \qquad (7.6)$$

Let us remark that (7.6) holds for all $t \ge 0$ such that $y(t)$ is defined. According to our condition (3), the right-hand side in (7.6) is bounded on R_+. This implies that a finite escape time cannot exist. Moreover, (7.6) and (7.2) show that $y(t)$ is bounded on R_+, i.e., the first inequality (7.5) is proved. Taking into account (7.2), one sees that $U(y)$ is bounded below on R^n. Therefore, (7.6) implies that $\dot{y} \in L^2$, and the second inequality (7.5) is thus established.

We shall now prove that the limit set of $y(t)$ agrees with that of a convenient solution of (7.4). Indeed, the family $\{y(t + \tau) : \tau \in R_+\}$ is (obviously) uniformly bounded and equicontinuous on R_+. The equicontinuity property follows from

$$\|y(s) - y(u)\| \le \left| \int_u^s \|\dot{y}(\tau)\| \, d\tau \right| \le c^{1/2} |s - u|^{1/2}.$$

Hence, there exists a sequence $\{\tau_n\}$, with $\tau_n \to \infty$, such that $y(t + \tau_n) \to z(t)$ uniformly on any compact interval of R_+. It will be shown further that $z(t)$ is a solution of (7.4).

First, let us prove that

$$\lim_{t \to \infty} \int_0^t k(t - \tau) \dot{y}(\tau) \, d\tau = 0. \tag{7.7}$$

Indeed, to any $\varepsilon > 0$ there corresponds $T = T(\varepsilon)$ with the property that

$$\int_T^\infty \|\dot{y}(\tau)\|^2 \, d\tau < \varepsilon^2 \left(\int_0^\infty \|k(t)\|^2 \, dt \right)^{-1}.$$

Consequently, for $t > T$

$$\left\| \int_0^t k(t - \tau) \dot{y}(\tau) \, d\tau \right\| \le \int_0^T \|k(t - \tau) \dot{y}(\tau)\| \, d\tau$$

$$+ \left(\int_T^t \|k(t - \tau)\|^2 \, d\tau \right)^{1/2} \left(\int_T^t \|\dot{y}(\tau)\|^2 \, d\tau \right)^{1/2}$$

$$\le \left(\int_{t-T}^\infty \|k(u)\|^2 \, du \right)^{1/2} c_1^{1/2} + \varepsilon.$$

If T is fixed, we can choose t so large that the first term in the last member becomes less than ε. This proves (7.7).

Taking into account that $\lambda(t) \in C_0$, we see that, formally, (7.1) leads to (7.4). Since the limit function $z(t)$ is not known to be differentiable, we have to furnish supplementary arguments. Let us denote by Δ the interval $(t, t + h)$, $h > 0$. From (7.1) we obtain by integration

$$\int_\Delta [\dot{y}(\tau + \tau_m) + K\dot{y}\,|_{\tau + \tau_m}] \, d\tau = -\int_\Delta \varphi(y_m(\tau)) \, d\tau + \int_\Delta \lambda(\tau + \tau_m) \, d\tau,$$

where $y_m(\tau) = y(\tau + \tau_m)$. Since $y_m(t) \to z(t)$ uniformly on any compact interval, we find

$$z(t + h) - z(t) = -\int_\Delta \varphi(z(\tau)) \, d\tau \tag{7.8}$$

taking into account (7.7). From (7.8) there results that $z(t)$ satisfies (7.4), $h > 0$ being arbitrary.

If we denote by Y the limit set of $y(t)$ and by Z that of $z(t)$, then $Z \subset Y$. Indeed, if $\zeta \in Z$, there exists a sequence $\{t_k\}$, $t_k \to \infty$, such that $z(t_k) \to \zeta$. But

$z(t_k) = \lim_{m\to\infty} y(t_k + \tau_m)$ and this yields $\zeta = \lim_{k\to\infty} y(t_k + \tau_{m_k})$, for a convenient subsequence $\{\tau_{m_k}\}$. Consequently, $\zeta \in Y$.

Conversely, if $\eta \in Y$, there exists a sequence $\{t_k\}$, $t_k \to \infty$, for which $y(t_k) \to \eta$. Let $z(t)$ be the solution of (7.4) with $z(0) = \eta$. It follows easily that $y(t + t_k) \to z(t)$ as $k \to \infty$, uniformly on any compact interval. If we chose a subsequence $\{t_{k_m}\} \subset \{t_k\}$ such that $t_{k_m} > 2t_m$, then for $t = t_{k_m} - t_m$ we have

$$\|y(t + t_m) - z(t)\| = \|y(t_{k_m}) - z(t_{k_m} - t_m)\| \to 0$$

as $m \to \infty$. But $t_{k_m} - t_m > t_m \to \infty$ as $m \to \infty$, which shows that

$$\eta = \lim_{m\to\infty} z(t_{k_m} - t_m) \in Z.$$

Therefore, $Y = Z$ and Theorem 7.1 is completely proven.

Corollary Assume that all conditions of Theorem 7.1 hold and, moreover, $U(y)$ has only a finite number of critical points (i.e., such that $\varphi(y) = 0$). Then any solution $y(t)$ of (7.1) approaches a critical point of $U(y)$ as $t \to \infty$.

Indeed, any solution of (7.4) approaches a singular point as $t \to \infty$. Let $z(t)$ be the solution of (7.4) such that $z(0) = z^0$. Then

$$\int_0^t \|\dot{z}(\tau)\|^2 \, d\tau + U(z(t)) \leq U(z^0)$$

for any $t \geq 0$ such that $z(t)$ exists. From the preceding inequality there results first that $z(t)$ is defined on R_+ and, moreover, that it is bounded on R_+. Since $\dot{z}(t)$ is also bounded on R_+, we get the uniform continuity of $z(t)$ on R_+. Again taking (7.4) into account, we see that $\dot{z}(t)$ is uniformly continuous. But $\int_0^t \|\dot{z}(\tau)\|^2 \, d\tau$ is bounded on R_+, and this implies $\dot{z}(t) \to 0$ as $t \to \infty$. In other words, for any solution $z(t)$ of (7.4) we have $\varphi(z(t)) \to 0$ as $t \to \infty$. There remains to show that this property implies $z(t) \to \zeta$, with $\varphi(\zeta) = 0$, as $t \to \infty$. From the boundedness of $z(t)$ on R_+, there results that its limit set is nonempty. Let ξ be a limit point of $z(t)$: $\xi = \lim_{m\to\infty} z(t_m)$, with $t_m \to \infty$. Then $\varphi(\xi) = 0$. But the limit set is connected, and according to our hypothesis on $U(y)$ there results $\lim_{t\to\infty} z(t) = \xi$. The corollary is thus proven.

3.8   Existence of L^2 Solutions

The frequency conditions can be used in order to ensure various kinds of behavior for the solutions of integral equations. For instance, we derived in Section 3.6, as a partial result, the fact that the solution belongs to L^2. We are now going to establish an existence result in the space L^2, for the integral equation

$$\sigma(t) = h(t) + \int_0^t k(t - s)f(s, \sigma(s)) \, ds, \tag{8.1}$$

where all the functions involved are scalar. The existence result we shall prove is due to Barbu [1]. The vector case was investigated by Halanay [3].

Theorem 8.1   Assume that the following conditions hold for Eq. (8.1):

1. $h(t) \in L^2(R_+, R)$;
2. $k(t) \in L(R_+, R) \cap L^2(R_+, R)$;
3. $f(t, \sigma)$ is continuous in σ and measurable in t for $t \in R_+$, $\sigma \in R$, and

$$|f(t, \sigma)| \leq L|\sigma| + c(t), \tag{8.2}$$

with $L > 0$ and $c(t) \in L^2(R_+, R)$;
4. there exists a real number q, $|q| < L^{-1}$, such that

$$\text{Re}\{|\tilde{k}(s)|^2 + q\tilde{k}(s)\} \leq 0, \qquad s \in R. \tag{8.3}$$

Then, there exists at least one solution $\sigma(t)$ of Eq. (8.1) such that

$$\sigma(t) \in L^2(R_+, R). \tag{8.4}$$

Proof   We shall apply the fixed-point theorem of Schauder and Tychonoff in the space $L^2_{\text{loc}}(R_+, R)$. This is a locally convex space whose elements are measurable functions from R_+ into R such that $\sigma \in L^2_{\text{loc}}$ if and only if $\sigma(t)$ is square integrable on any compact interval of R_+. The topology of $L^2_{\text{loc}}(R_+, R)$ is given by the family of seminorms

$$|\sigma|_n = \left\{ \int_0^n \sigma^2(t) \, dt \right\}^{1/2}, \qquad n = 1, 2, \dots.$$

The space $L^2_{\text{loc}}(R_+, R)$ is a Fréchet space.

Let $\mathcal{M}$ be the set of all functions belonging to $L^2_{\text{loc}}(R_+, R)$ such that

$$\int_0^\infty \sigma^2(t) \, dt \leq M^2, \tag{8.5}$$

where M is a positive number satisfying the inequality

$$M > (c_0|q| + h_0)/(1 - L|q|), \tag{8.6}$$

with $c_0 = |c|_{L^2}$ and $h_0 = |h|_{L^2}$. This set is convex and closed in $L^2_{\text{loc}}(R_+, R)$.
We shall now consider on $\mathcal{M}$ the operator T given by

$$(T\sigma)(t) = h(t) + \int_0^t k(t - s)f(s, \sigma(s)) \, ds. \tag{8.7}$$

First, it can be proved that T is continuous from $L^2_{\text{loc}}(R_+, R)$ into itself. Indeed, the convergence of a sequence $\{\sigma_m\} \subset L^2_{\text{loc}}$ means that this sequence converges in $L^2([0, n], R)$ for any n. If we denote also by σ the restriction of this function to $[0, n]$ for fixed n, then the mapping $\sigma \to f(\cdot, \sigma)$ from $L^2([0, n], R)$

into itself is continuous (see, for instance, Krasnoselski [1]). This remark leads easily to the conclusion that for any fixed n, the mapping $\sigma \to T\sigma$ is continuous from $L^2([0, n], R)$ into itself. Therefore, the mapping $\sigma \to T\sigma$ is continuous from $L^2_{\text{loc}}(R_+, R)$ into itself.

Second, let us show that

$$T\mathcal{M} \subset \mathcal{M}. \tag{8.8}$$

In order to prove (8.8) we shall use the truncation and frequency techniques. For any $t > 0$, we define

$$f_t(\tau) = \begin{cases} f(\tau, \sigma(\tau)), & 0 \le \tau \le t, \\ 0, & \tau > t, \end{cases}$$

with $\sigma \in \mathcal{M}$. Let us denote

$$\kappa_t(\tau) = \int_0^\tau k(\tau - s) f_t(s) \, ds. \tag{8.9}$$

Under conditions (2) and (3) from our theorem, there results

$$f_t, \kappa_t \in L(R_+, R) \cap L^2(R_+, R). \tag{8.10}$$

From Parseval's equality one obtains

$$\int_0^\infty \kappa_t(\tau)[\kappa_t(\tau) + q f_t(\tau)] \, d\tau = 2\pi^{-1} \int_{-\infty}^\infty \text{Re}\{\tilde{\kappa}_t(s)[\overline{\tilde{\kappa}_t(s) + q \tilde{f}_t(s)}]\} \, ds.$$

Taking into account (8.9), the preceding equality can be written in the form:

$$\int_0^\infty \kappa_t(\tau)[\kappa_t(\tau) + q f_t(\tau)] \, d\tau = 2\pi^{-1} \int_{-\infty}^\infty \text{Re}\{|\tilde{k}(s)|^2 + q\tilde{k}(s)\} |\tilde{f}_t(s)|^2 \, ds. \tag{8.11}$$

According to our assumption (4), from (8.11) we derive

$$\int_0^\infty \kappa_t(\tau)[\kappa_t(\tau) + q f_t(\tau)] \, d\tau \le 0 \tag{8.12}$$

for any $t > 0$. Taking into account condition (3), we obtain from (8.12) the following inequality:

$$\int_0^\infty \kappa_t^2(\tau) \, d\tau \le |q| \left\{ \int_0^\infty \kappa_t^2(\tau) \, d\tau \right\}^{1/2} \left\{ \int_0^\infty f_t^2(\tau) \, d\tau \right\}^{1/2}.$$

This gives

$$\left\{ \int_0^\infty \kappa_t^2(\tau) \, d\tau \right\}^{1/2} \le |q|(LM + c_0), \tag{8.13}$$

if we consider the definition of f_t and condition (3) from the statement of the theorem. Since $\kappa_t(\tau) = \int_0^\tau k(\tau - s)f(s, \sigma(s))\,ds = \kappa(\tau)$ for $\tau < t$, from (8.13) and $(T\sigma)(t) = h(t) + \kappa(t)$, we obtain

$$\left\{ \int_0^\infty |(T\sigma)(t)|^2\,dt \right\}^{1/2} \leq h_0 + |q|(LM + c_0) \leq M,$$

i.e., $T\sigma \in \mathcal{M}$ for $\sigma \in \mathcal{M}$.

We shall now prove that the set $T\mathcal{M}$ is relatively compact in $L^2_{\text{loc}}(R_+, R)$. Let us fix a positive integer n and remark that

$$\left\{ \int_0^n |(T\sigma)(t + \tau) - (T\sigma)(t)|^2\,dt \right\}^{1/2}$$

$$\leq \left\{ \int_0^n |h(t + \tau) - h(t)|^2\,dt \right\}^{1/2}$$

$$+ \left\{ \int_0^n dt \left[\int_0^t |k(t + \tau - s) - k(t - s)|\,|f(s, \sigma(s))|\,ds \right]^2 \right\}^{1/2}$$

$$+ \left\{ \int_0^n dt \left[\int_t^{t+\tau} |k(t + \tau - s)|\,|f(s, \sigma(s))|\,ds \right]^2 \right\}^{1/2}.$$

By means of the Schwartz inequality applied in the last two integrals, we get

$$\left\{ \int_0^n |(T\sigma)(t + \tau) - (T\sigma)(t)|^2\,dt \right\}^{1/2} \leq \omega_0(\tau) + K[\omega_1(\tau) + \omega_2(\tau)],$$

where

$$\omega_0(\tau) = \left\{ \int_0^n |h(t + \tau) - h(t)|^2\,dt \right\}^{1/2},$$

$$\omega_1(\tau) = \left\{ \int_0^n dt \int_0^t |k(t + \tau - s) - k(t - s)|^2\,ds \right\}^{1/2},$$

$$\omega_2(\tau) = \left\{ \int_0^n dt \int_t^{t+\tau} |k(t + \tau - s)|^2\,ds \right\}^{1/2},$$

$$K = \left\{ \int_0^\infty f^2(s, \sigma(s))\,ds \right\}^{1/2}.$$

Since $\omega_0(\tau)$, $\omega_1(\tau)$, and $\omega_2(\tau)$ are obviously continuous at $\tau = 0$, there results that the functions belonging to $T\mathcal{M}$ are equicontinuous in the mean (of order 2), on any compact interval of R_+. From the inclusion (8.8) there results that $T\mathcal{M}$ is bounded on any compact interval. Therefore, the set $T\mathcal{M}$ is relatively compact in $L^2_{\text{loc}}(R_+, R)$ (see, for instance, Yosida [1]).

From the preceding considerations we see that the fixed-point theorem of Schauder and Tychonoff can be applied to the operator T. Therefore, Theorem 8.1 is proved.

Remark 1 If we assume that Eq. (8.1) satisfies, besides the conditions of Theorem 8.1, also

$$h'(t),\ k'(t) \in L(R_+, R), \tag{8.14}$$

then from

$$\sigma'(t) = h'(t) + k(0)f(t, \sigma(t)) + \int_0^t k'(t - s)f(s, \sigma(s))\ ds$$

we see that $\sigma(t)$ is uniformly continuous on R_+. Since $\sigma \in L^2$, we obtain that $\sigma \in C_0(R_+, R)$. In other words, the frequency condition (8.3) also suffices to ensure the classical stability condition.

Remark 2 It appears interesting to compare the frequency condition (8.3) from Theorem 8.1 with Popov's frequency condition

$$\mathrm{Re}\{(1 - isq)\tilde{k}(s)\} - L^{-1} \le 0, \qquad s \in R, \tag{8.15}$$

which ensures the asymptotic stability of Eq. (1.1), under conditions stated in Theorem 2.2 except that condition (2.8) should be replaced by

$$0 < \sigma\varphi(\sigma) < L\sigma^2$$

for any $\sigma \in R$ (see Albertoni and Szegö [1]).

Condition (8.3) states that $|\tilde{k}(s)|^2 + q\ \mathrm{Re}\ \tilde{k}(s) \le 0$, from which we derive $|\tilde{k}(s)|^2 \le |q|\ \mathrm{Re}\ \tilde{k}(s)$ and $|\mathrm{Re}\ \tilde{k}(s)| \le |q| < L^{-1}$. This shows that

$$\mathrm{Re}\ \tilde{k}(s) - L^{-1} < 0, \qquad s \in R. \tag{8.16}$$

Consequently, condition (8.3) implies (8.15) with $q = 0$.

As pointed out by Halanay [3], this is quite natural if we take into account that condition (8.3) ensures—in fact—not only stability, but also that σ belongs to L^2.

3.9 Control Systems with Time Lag

As shown several times in the preceding sections, the results obtained for integral equations can be easily applied to the stability problem of differential systems of the form (1.4) or (1.8). The aim of this section is to illustrate another significant application of the above results. Namely, as pointed out by Halanay [1, 4], the stability problem of automatic control systems with time lag can be also reduced to the investigation of an integral equation.

Let us consider the system

$$\dot{x}(t) = Ax(t) + Bx(t - h) + c\varphi(\sigma), \qquad \sigma = (d, x), \tag{9.1}$$

where x is an n-vector function, A and B are square constant matrices of type n by n, $\varphi(\sigma)$ is a scalar function, and c, d are constant n-vectors. By assumption, all the quantities occurring in (9.1) are real. Obviously, we have to assume $h > 0$.

If we assume further that $\varphi(\sigma)$ is a continuous function for $\sigma \in R$, then a local existence theorem holds for (9.1). More precisely, for any $t_0 \in R$ and $x_0(t) \in C([t_0 - h, t_0], R^n)$, there exists at least one solution $x = x(t)$ of (9.1), defined on $[t_0 - h, T)$, $T > t_0$, such that the initial condition $x(t) = x_0(t)$, $t \in [t_0 - h, t_0]$, be verified. Let us remark that richer function spaces than $C([t_0 - h, t_0], R^n)$ can be used.

A variation-of-constants formula holds for any system of the form

$$\dot{x}(t) = Ax(t) + Bx(t - h) + f(t). \tag{9.2}$$

It states that

$$x(t) = X(t - t_0)x(t_0) + \int_{t_0-h}^{t_0} X(t - u - h)Bx(u)\, du + \int_{t_0}^{t} X(t - u)\, f(u)\, du.$$

$$\tag{9.3}$$

$X(t)$ is determined by $\dot{X}(t) = AX(t) + BX(t - h)$, $X(0) = I$, $X(t) = 0$ for $t < 0$. The validity of (9.3) is proven, for instance, in Halanay's book [4]. He considers a more general case when A and B are periodic functions of t.

From (9.1) we can now derive an integral equation for $\sigma = \sigma(t)$. Indeed, by means of the variation-of-constants formula we get (with $t_0 = 0$) from the first Eq. (9.1):

$$x(t) = X(t)x^0 + \int_{-h}^{0} X(t - u - h)Bx(u)\, du + \int_{0}^{t} X(t - u)c\varphi(\sigma(u))\, du. \tag{9.4}$$

If we substitute $x(t)$ from (9.4) in the second Eq. (9.1), we obtain

$$\sigma(t) = \left(d,\, X(t)x^0 + \int_{-h}^{0} X(t - u - h)Bx(u)\, du \right) + \int_{0}^{t} (d,\, X(t - u)c)\varphi(\sigma(u))\, du,$$

$$\tag{9.5}$$

i.e., $\sigma(t)$ satisfies an integral equation of the form

$$\sigma(t) = h(t) + \int_{0}^{t} k(t - u)\varphi(\sigma(u))\, du, \tag{9.6}$$

with

$$h(t) = \left(d,\, X(t)x^0 + \int_{-h}^{0} X(t - u - h)Bx(u)\, du \right),$$

$$k(t) = (d,\, X(t)c).$$

The characteristic equation associated with the homogeneous system $\dot{x}(t) = Ax(t) + Bx(t - h)$ is

$$\det(A + Be^{-\lambda h} - \lambda I) = 0. \tag{9.7}$$

A basic result of Hale [1] states that there exists a positive constant K such that

$$\|X(t)\| \le Ke^{-\alpha t}, \qquad t \ge 0, \tag{9.8}$$

as long as any root of (9.7) satisfies $\operatorname{Re} \lambda \le -\alpha$. Hence, $X(t)$ has exponential decay at infinity if all the characteristic roots of (9.7) lie in a half-plane $\operatorname{Re} \lambda \le -\alpha < 0$.

Consequently, if we assume that there exists $\alpha > 0$ such that

$$\operatorname{Re} \lambda \le -\alpha \tag{9.9}$$

for any root of (9.7), the functions $h(t)$ and $k(t)$ from Eq. (9.6) have exponential decay at infinity.

Before stating the stability result concerning the system (9.1), we shall find the Fourier transform of the fundamental matrix $X(t)$. Taking into account its definition we obtain

$$-is\tilde{X}(s) - I = A\tilde{X}(s) + e^{ish}B\tilde{X}(s),$$

whence

$$\tilde{X}(s) = -(A + e^{ish}B + isI)^{-1}. \tag{9.10}$$

Let us remark that the inverse matrix exists for any real s, because $\lambda = -is$ cannot be a root of the characteristic equation (9.7).

Theorem 9.1 Assume that the following conditions hold for the system (9.1):

1. there exists a positive α such that (9.9) is satisfied for any root λ of Eq. (9.7);
2. $\varphi(\sigma)$ is continuous and bounded from R into itself, and such that $\sigma\varphi(\sigma) > 0$ for $\sigma \ne 0$;
3. there exists $q \ge 0$ such that the frequency condition

$$\operatorname{Re}\{(isq - 1)(d, (A + e^{ish}B + isI)^{-1}c)\} \le 0 \tag{9.11}$$

is satisfied for any $s \in R$.

Then any solution $x(t)$ of system (9.1) approaches zero as $t \to \infty$.

Proof As seen above, $\sigma = \sigma(t)$ satisfies an integral equation (9.5) or, equivalently, (9.6). This [Eq. (9.6)] is an equation satisfying all the conditions required for the application of Theorem 2.2. Hence $\sigma(t) \to 0$ as $t \to \infty$. Moreover, according to formula (9.4), we obtain $x(t) \to 0$ as $t \to \infty$, no matter how we choose the initial function $x_0(t)$, $t \in [-h, 0]$.

In other words, the absolute stability of the control system with time lag (9.1) is assured under condition (9.9) for the characteristic equation and the frequency condition (9.11).

3.10 A Result of Yakubovitch

We shall now discuss the vector integral equation

$$\sigma(t) = h(t) + \int_0^t k(t - s)\varphi(s, \sigma(s))\, ds, \tag{10.1}$$

where σ, $h \in R^n$, $\varphi \in R^m$, and $k(t)$ is a matrix kernel of type n by m. Equation (10.1) arises, for instance, in studying multiple-input multiple-output feedback systems.

The result we shall establish is due to Yakubovitch [1]. His new idea was to consider some quadratic connections between the components of φ and σ, generalizing such conditions as $\sigma\varphi(\sigma) > 0$ ($\sigma \neq 0$) from Theorems 2.2 and 3.1, or condition (6.2) from Theorem 6.1. In the more general setting of Yakubovitch, the problem of finding frequency conditions of stability can be also solved by using Popov's technique.

The existence problem will not be discussed in this section. We shall assume that there exists at least one solution of (10.1) which is defined on the positive half-axis. The main attention is paid to the behavior of solutions.

Before stating the main result of this section we need some auxiliary considerations.

We will assume that for any fixed $t \in R_+$, φ satisfies the following relations:

$$F_j(\varphi, \sigma) = 0, \qquad j = 1, 2, \ldots, p, \tag{10.2}$$

and

$$F_j(\varphi, \sigma) \geq 0, \qquad j = p + 1, \ldots, p + q, \tag{10.3}$$

where $F_j(\varphi, \sigma)$, $j = 1, 2, \ldots, p + q$, are real quadratic forms in φ_1, φ_2, $\ldots$, φ_m, σ_1, σ_2, $\ldots$, σ_n. With respect to p and q, we assume that $p \geq 0$ and $q > 0$.

Let us consider an example. When $m = n$ and $\varphi_j(\sigma) = \varphi_j(\sigma_j)$, $j = 1, 2, \ldots$, n, it is usually assumed that

$$0 \leq \varphi_j(\sigma_j)/\sigma_j \leq \mu_j, \qquad j = 1, 2, \ldots, n, \tag{10.4}$$

for any $\sigma_j \neq 0$. If we denote

$$F_j(\varphi, \sigma) = \varphi_j(\sigma_j - \mu_j^{-1}\varphi_j) \qquad j = 1, 2, \ldots, n, \tag{10.5}$$

then (10.4) leads to

$$F_j(\varphi, \sigma) \geq 0, \qquad j = 1, 2, \ldots, n. \tag{10.6}$$

In other words, conditions (10.2) and (10.3) are verified with $p = 0, q = n = m$.
Consider now the quadratic form

$$F(\varphi, \sigma) = \sum_{j=1}^{p+q} \tau_j F_j(\varphi, \sigma), \qquad (10.7)$$

where $\tau_j \geq 0$, $j = p + 1, \ldots, p + q$, and τ_j are arbitrary real numbers for
$j = 1, 2, \ldots, p$. Let us extend this real quadratic form to a complex hermitian
form. This means that any product $\xi\eta$ should be replaced by $\mathrm{Re}(\xi\bar{\eta})$, $\bar{\eta}$ being
the conjugate of η. In particular, ξ^2 should be replaced by $|\xi|^2$. For instance,
instead of (10.5), we have to consider $F_j(\varphi, \sigma) = \mathrm{Re}(\sigma_j \bar{\varphi}_j) - \mu_j^{-1} |\varphi_j|^2$.

With the hermitian form $F(\varphi, \sigma)$, as previously defined, we shall associate
a family of hermitian forms $\tilde{F}(s, \tilde{\varphi})$, $s \in R$, $\tilde{\varphi}$ being a complex n-vector. These
forms are given by

$$\tilde{F}(s, \tilde{\varphi}) = F(\tilde{\varphi}, \tilde{k}(s)\tilde{\varphi}), \qquad (10.8)$$

with

$$\tilde{k}(s) = \int_0^\infty k(t)e^{ist}\, dt. \qquad (10.9)$$

Of course, we have to assume that (10.9) makes a sense, for instance, that
$\|k\| \in L(R_+, R)$.

We are now able to state the following result.

Theorem 10.1 Consider the integral equation (10.1) and let $\sigma(t)$ be a
measurable solution defined on R_+. Assume that the following conditions
hold:

1. $h(t) \in L^2(R_+, R^n)$;
2. $k(t)$ is a matrix kernel of type n by m such that $\|k(t)\| \in L(R_+, R)$;
3. $\varphi(t, \sigma(t)) \in L^2_{\mathrm{loc}}(R_+, R^m)$, where $\sigma(t)$ denotes the solution of (10.1);
4. φ is subject to conditions (10.2) and (10.3);
5. there exist real numbers τ_j, $j = 1, 2, \ldots, p + q$, with $\tau_j \geq 0$ for
$j = p + 1, \ldots, p + q$, such that $F(\varphi, \sigma)$ given by (10.7) satisfies

$$F(0, \sigma) \geq 0 \qquad \text{for any} \quad \sigma \in R^n; \qquad (10.10)$$

6. the quadratic form $\tilde{F}(s, \tilde{\varphi})$ is such that there exists $\lambda > 0$ with

$$\tilde{F}(s, \tilde{\varphi}) \leq -\lambda \|\tilde{\varphi}\|^2, \qquad \tilde{\varphi} \in R^m, \qquad (10.11)$$

for any $s \in R$.

Then, the solution $\sigma(t)$ satisfies

$$\sigma(t) \in L^2(R_+, R^n), \qquad \varphi(t, \sigma(t)) \in L^2(R_+, R^m). \qquad (10.12)$$

Proof From condition (2), there results that $\|\tilde{k}(s)\|$ is bounded on R_+. Let $M > 0$ be such that

$$M = \sup_{s \in R}(1 + \|\tilde{k}(s)\|^2). \tag{10.13}$$

According to condition (6), we can find a positive δ such that for any $s \in R$

$$\tilde{F}(s, \tilde{\varphi}) \leq -2M\delta\|\tilde{\varphi}\|^2, \qquad \tilde{\varphi} \in R^m. \tag{10.14}$$

Let us consider now the following two hermitian forms $G_j(\varphi, \sigma)$, $j = 1, 2$, given by

$$G_j(\varphi, \sigma) = F(\varphi, \sigma) + j\delta(\|\varphi\|^2 + \|\sigma\|^2), \tag{10.15}$$

where $F(\varphi, \sigma)$ is defined by (10.7).

Let $t > 0$ be an arbitrary number and consider the functions σ_t and φ_t obtained by truncation procedure from σ and φ, respectively. In other words, $\sigma_t(\tau) = \sigma(\tau)$ for $0 \leq \tau \leq t$ and $\sigma_t(\tau) = 0$ for $\tau > t$. Analogously, $\varphi_t(\tau) = \varphi(\tau, \sigma(\tau))$ for $0 \leq \tau \leq t$ and $\varphi_t(\tau) = 0$ for $\tau > t$.

By assumption, $\sigma(t)$ is a solution of (10.1), with φ subject to (10.2) and (10.3). Therefore, $F(\varphi(t, \sigma(t)), \sigma(t)) \geq 0$ for any $t > 0$. Consequently, we can write

$$\delta \int_0^t (\|\varphi_t\|^2 + \|\sigma_t\|^2)\, d\tau \leq \int_0^t G_1(\varphi_t, \sigma_t)\, d\tau, \tag{10.16}$$

both integrals occurring in (10.16) being convergent. Indeed, $\varphi_t \in L^2(R_+, R^m)$ according to condition (3) from Theorem 10.1. If we define (for any $t > 0$)

$$\xi_t(\tau) = \int_0^\tau k(\tau - s)\varphi_t(s)\, ds, \qquad \tau \in R_+, \tag{10.17}$$

then Eq. (10.1) shows that $\sigma_t = h_t + \xi_t$ on $[0, t]$. Conditions (2) and (3) of Theorem 10.1 yield $\xi_t \in L^2(R_+, R^n)$. Hence, $\sigma_t \in L^2(R_+, R^n)$ and this shows that both integrals in (10.16) make a sense.

Let us now represent the hermitian form $G_1(\varphi, \sigma)$ as a sum, in the following way:

$$G_1(\varphi, \sigma) = (\varphi, A\varphi) + 2(\varphi, B\sigma) + (\sigma, C\sigma), \tag{10.18}$$

where A, B, and C are real constant matrices. A and C are m by m and n by n symmetric matrices, respectively, while B is of type n by m.

Since $\sigma_t = h_t + \xi_t$, we obtain

$$G_1(\varphi_t, \sigma_t) = G_1(\varphi_t, \xi_t) + 2(\varphi_t, Bh_t) + 2(\xi_t, Ch_t) + (h_t, Ch_t). \tag{10.19}$$

Taking into account that $2|(u, v)| \leq \delta\|u\|^2 + \delta^{-1}\|v\|^2$ for any pair of n-vectors u, v, one obtains

$$2|(\varphi_t, Bh_t)| \le \delta\|\varphi_t\|^2 + \delta^{-1}\|B\|^2\|h_t\|^2,$$

$$2|(\xi_t, Ch_t)| \le \delta\|\xi_t\|^2 + \delta^{-1}\|C\|^2\|h_t\|^2.$$

From (10.19) and the above inequalities there results

$$G_1(\varphi_t, \sigma_t) \le G_2(\varphi_t, \xi_t) + K\|h_t\|^2, \tag{10.20}$$

where

$$K = \|C\| + \delta^{-1}(\|B\|^2 + \|C\|^2). \tag{10.21}$$

From (10.16) and (10.20) we get

$$\delta \int_0^t (\|\varphi_t\|^2 + \|\sigma_t\|^2)\, d\tau \le \int_0^t G_2(\varphi_t, \xi_t)\, d\tau + K\|h\|_{L^2}. \tag{10.22}$$

In order to end the proof of Theorem 10.1, it suffices to show that the integral in the right-hand side of (10.22) is ≤ 0. Indeed, let us consider the quantity

$$I(t) = \int_0^\infty G_2(\varphi_t, \xi_t)\, d\tau. \tag{10.23}$$

This is finite for any $t \in R_+$, because both φ_t and ξ_t belong to L^2. Parseval's equality gives

$$I(t) = (2\pi)^{-1} \int_R G_2(\tilde{\varphi}_t, \tilde{\xi}_t)\, ds. \tag{10.24}$$

Formula (10.15) yields for $j = 2$:

$$G_2(\tilde{\varphi}_t, \tilde{\xi}_t) = \tilde{F}(s, \tilde{\varphi}_t) + 2\delta(\|\tilde{\varphi}_t\|^2 + \|\tilde{\xi}_t\|^2), \tag{10.25}$$

taking into account (10.8) and (10.17). We get further

$$G_2(\tilde{\varphi}_t, \tilde{\xi}_t) \le -2M\delta\|\tilde{\varphi}_t\|^2 + 2M\delta\|\tilde{\varphi}_t\|^2 = 0,$$

considering (10.13), (10.14), and (10.17). The last inequality implies $I(t) \le 0$, $t \in R_+$. On the other hand, (10.23) can be written in the form

$$I(t) = \int_0^t G_2(\varphi_t, \xi_t)\, d\tau + \int_t^\infty G_2(0, \xi_t)\, d\tau,$$

inasmuch as $\varphi_t(\tau) = 0$ for $\tau > t$. But $G_2(0, \xi_t) \ge 0$, according to (10.10) and (10.15). Therefore, $\int_0^t G_2(\varphi_t, \xi_t)\, d\tau \le 0$, for any $t \in R_+$. Theorem 10.1 is completely proved.

Remark 1   From (10.22) and the fact that the integral in the right member of this formula is nonpositive, there results

$$\delta \int_0^\infty (\|\varphi(t, \sigma(t))\|^2 + \|\sigma(t)\|^2)\, dt \le K\|h\|_{L^2}. \tag{10.26}$$

If we make further assumptions on $\sigma(t)$, for instance that it is uniformly continuous on R_+, then we get a better result: $\sigma(t) \in L^2 \cap C_0$.

Remark 2   The result obtained in Theorem 10.1 remains true if conditions (10.3) are replaced by

$$\int_0^t F_j(\varphi, \sigma) \, d\tau \geq -\gamma_j, \qquad j = p+1, \ldots, p+q, \tag{10.27}$$

where γ_j are positive constants. Indeed, instead of inequality (10.16), we obtain in this case

$$\delta \int_0^t (\|\varphi_t\|^2 + \|\sigma_t\|^2) \, d\tau \leq \int_0^t G_1(\varphi_t, \sigma_t) \, d\tau + \sum_{p+1}^{p+q} \tau_j \gamma_j.$$

Afterwards, the proof follows the same lines as above. It is obvious that conditions (10.27) are more general than (10.3).

Remark 3   It is possible to obtain an analogous result for the solution of the equation

$$\sigma(t) = h(t) + \int_0^t k(t-s)\varphi(s, \sigma(s)) \, ds + R\varphi(t, \sigma(t)), \tag{10.28}$$

with R a constant matrix of type n by m.

Before concluding this section, we shall consider another kind of connection between φ and σ. Namely, instead of (10.2) and (10.3), we assume that

$$F_j(\varphi, \sigma, \sigma') = 0, \qquad j = 1, 2, \ldots, p, \tag{10.29}$$

and

$$F_j(\varphi, \sigma, \sigma') \geq 0, \qquad j = p+1, \ldots, p+q. \tag{10.30}$$

This time, the F_j are real quadratic forms of their arguments and the prime denotes the derivative.

Define now the quadratic form $F(\varphi, \sigma, \sigma')$ by means of the formula

$$F(\varphi, \sigma, \sigma') = \sum_{j=1}^{p+q} \tau_j F_j(\varphi, \sigma, \sigma'), \tag{10.31}$$

where $\tau_j, j = 1, 2, \ldots, p$, are arbitrary real numbers and $\tau_j \geq 0, j = p+1, \ldots, p+q$. Let $\tilde{F}(s, \tilde{\varphi})$ be the hermitian form in $\tilde{\varphi}$, for any $s \in R$, defined by

$$\tilde{F}(s, \tilde{\varphi}) = F(\tilde{\varphi}, \tilde{k}(s)\tilde{\varphi}, -is\tilde{k}(s)\tilde{\varphi}). \tag{10.32}$$

The following result can be easily obtained from Theorem 10.1 (more precisely, from the analogous statement concerning Eq. (10.28)).

Theorem 10.2   Consider Eq. (10.1) under the following assumptions:

1. $h(t)$, $h'(t) \in L^2(R_+, R^n)$;
2. $k(t)$ is a matrix kernel of type n by m such that

$$\|k(t)\|, \ \|k'(t)\| \in L(R_+, R);$$

3. there exists a measurable solution $\sigma(t)$ on R_+ such that

$$\varphi(t, \sigma(t)) \in L^2_{\text{loc}}(R_+, R^m);$$

4. conditions (10.29) and (10.30) are verified;
5. the form $F(\varphi, \sigma, \sigma')$ given by (10.31) is such that

$$F(0, \sigma, \sigma') \geq 0; \tag{10.33}$$

6. there exists $\lambda > 0$ such that

$$\tilde{F}(s, \tilde{\varphi}) \leq -\lambda\|\tilde{\varphi}\|^2, \qquad \tilde{\varphi} \in R^m, \tag{10.34}$$

for any $s \in R$.

Then $\sigma(t)$ satisfies

$$\|\sigma(t)\|, \ \|\sigma'(t)\|, \ \|\varphi(t, \sigma(t))\| \in L^2(R_+, R). \tag{10.35}$$

The proof follows easily from the next consideration related to (10.1) and the equation obtained by differentiating both members of (10.1) with respect to t:

$$\sigma'(t) = h'(t) + k(0)\varphi(t, \sigma(t)) + \int_0^t k'(t-s)\varphi(s, \sigma(s))\, ds. \tag{10.36}$$

If we denote

$$\hat{\sigma} = \begin{pmatrix} \sigma \\ \sigma' \end{pmatrix}, \qquad \hat{h} = \begin{pmatrix} h \\ h' \end{pmatrix}, \qquad \hat{k} = \begin{pmatrix} k \\ k' \end{pmatrix}, \qquad R = \begin{pmatrix} 0 \\ k(0) \end{pmatrix},$$

then the system (10.1), (10.36) can be written as

$$\hat{\sigma}(t) = \hat{h}(t) + \int_0^t \hat{k}(t-s)\varphi(s, \hat{\sigma}(s))\, ds + R\varphi(t, \hat{\sigma}(t)),$$

where $\varphi(t, \hat{\sigma}(t)) = \varphi(t, \sigma(t))$ by definition. The last equation is of the form (10.28), and the conditions of Theorem 10.1 are verified.

3.11   Energetic Stability

The concept of energetic stability was introduced by Kudrewicz [1], [2]. He gave several results concerning integral equations of the form

$$\sigma(t) = h(t) + \int_0^t k(t-s)\varphi(s, \sigma(s))\, ds + a\varphi(t, \sigma(t)), \tag{11.1}$$

and, more generally he discussed the energetic stability of feedback systems described by the functional equation

$$\sigma = \Gamma(\sigma) + h, \tag{11.2}$$

where Γ is a certain operator.

Before formulating the definition of this new concept of stability, we shall introduce a space of measurable functions that plays the central role. Only the scalar case will be discussed below.

Let $\mathcal{M}_2 = \mathcal{M}_2(R_+, R)$ be the subset of $L^2_{loc}(R_+, R)$ consisting of all $\sigma(t)$ for which

$$\limsup_{T \to \infty}\left\{T^{-1} \int_0^T |\sigma(s)|^2 \, ds\right\} = |\sigma| < +\infty. \tag{11.3}$$

The mapping $\sigma \to |\sigma|$ from $\mathcal{M}_2$ into R_+ is a seminorm on $\mathcal{M}_2$. In other words, the following conditions hold true:

$$|\sigma| \geq 0, \qquad \text{for any } \sigma \in \mathcal{M}_2; \tag{11.4}$$

$$|c\sigma| = |c|\,|\sigma| \qquad \text{for any } c \in R \text{ and } \sigma \in \mathcal{M}_2; \tag{11.5}$$

$$|\sigma_1 + \sigma_2| \leq |\sigma_1| + |\sigma_2| \qquad \text{for any } \sigma_1, \sigma_2 \in \mathcal{M}_2. \tag{11.6}$$

That $|\sigma| = 0$ does not necessarily imply $\sigma = 0$ a.e. on R_+ can be easily seen if we observe that $\sigma \in C_0(R_+, R)$ leads to $|\sigma| = 0$. The task of checking the validity of (11.5) and (11.6) is left to the reader.

Let us remark further that the space $\mathcal{M}_2(R_+, \mathscr{C})$, where $\mathscr{C}$ stands for the complex number field, can be defined in the same manner as above.

Equation (11.2) is called energetically stable if for any $h \in \mathcal{M}_2$, with $|h| = 0$, there results $\sigma \in \mathcal{M}_2$ and $|\sigma| = 0$. In the theory of systems, one uses the term energetically stable for a feedback system governed by an equation of the form (11.2) that is energetically stable in the sense considered above.

It is clear that the concept of energetic stability is related to a certain operator—namely, the operator occurring in the equation of the form (11.2) that describes the system under consideration. In order to formulate some simple conditions of energetic stability, it appears useful to introduce the following quantity associated with Γ:

$$\|\Gamma\| = \sup\{|\Gamma\sigma| / |\sigma| : |\sigma| \neq 0\}. \tag{11.7}$$

We denote it by Γ because its definition reminds us of the way one defines the norm of a linear operator.

Lemma 11.1   If Γ is an operator from $\mathcal{M}_2$ into itself and $\|\Gamma\| < 1$, then (11.2) is energetically stable.

Indeed, if σ is a solution of (11.2) belonging to $\mathcal{M}_2$, then

$$|\sigma| \leq \|\Gamma\| \, |\sigma| + |h| = \|\Gamma\| \, |\sigma|$$

when $|h| = 0$. One then obtains $|\sigma| = 0$, because $\|\Gamma\| < 1$.

From Lemma 11.1, we can easily derive the following result.

Lemma 11.2 Consider Eq. (11.2) with $\Gamma = AF$, where A is a linear operator and F is nonlinear. Assume that there exists λ with the following properties:

$$\|(I + \lambda A)^{-1}\| < +\infty, \tag{11.8}$$

where I denotes the identity operator in $\mathcal{M}_2$;

$$\|(I + \lambda A)^{-1} A\| \, \|F + \lambda I\| < 1. \tag{11.9}$$

Then, the system governed by Eq. (11.2) is energetically stable.

Indeed, Eq. (11.2) can be written in the equivalent form

$$\sigma + \lambda A \sigma = A[F(\sigma) + \lambda \sigma] + h. \tag{11.10}$$

Since $I + \lambda A$ is an invertible operator in $\mathcal{M}_2$, we obtain from (11.10):

$$\sigma = (I + \lambda A)^{-1} A(F + \lambda I)(\sigma) + (I + \lambda A)^{-1} h. \tag{11.11}$$

Equation (11.11) is of the form (11.2), and the conditions of Lemma 11.1 are obviously verified.

Consider now Eq. (11.1) and note that it can be written in the form

$$\sigma = AF(\sigma) + h, \tag{11.12}$$

with

$$(Ax)(t) = \int_0^t k(t - s)x(s) \, ds + ax(t) \tag{11.13}$$

and

$$(F\sigma)(t) = \varphi(t, \sigma(t)). \tag{11.14}$$

We are going to establish a basic result concerning the operator A given by (11.13). More precisely, we are interested in finding the quantity $\|A\|$.

In order to simplify the proofs, we shall consider L^∞ as underlying space. We have

$$L^\infty \subset \mathcal{M}_2. \tag{11.15}$$

In other words, we are concerned only with essentially bounded functions. Accordingly, in defining $\|A\|$ we consider only functions from L^∞.

Theorem 11.1 Let $k(t)$ be a locally integrable (on R_+) kernel such that

$$|k(t)| \leq Mt^{-(1+\varepsilon)} \qquad \text{a.e.,} \tag{11.16}$$

for sufficiently large t, where M and ε are some positive constants. If a is an arbitrary number, then A acts from L^∞ into itself and

$$\|A\| = \sup\{|K(s)| : s \in R\}, \tag{11.17}$$

where $K(s) = a + \tilde{k}(s)$, $\tilde{k}(s)$ is the Fourier transform of $k(t)$:

$$\tilde{k}(s) = \int_0^\infty k(t)e^{ist}\, dt, \qquad s \in R. \tag{11.18}$$

Proof Let us remark first that $\tilde{k}(s)$ has a meaning. Indeed, from (11.16) there results that $|k| \in L(R_+, R)$. Consequently, A carries L^∞ into itself. We denote

$$A_0 = \sup_{s \in R} |K(s)|. \tag{11.19}$$

We shall prove first that for any $x \in L^\infty$, there results $Ax \in L^\infty$ and

$$|Ax| \leq A_0|x|, \tag{11.20}$$

where A_0 is given by (11.19).

For any $x \in L^2_{\text{loc}}$ and $T > 0$, let us denote

$$|x|_T = \left\{ T^{-1}\int_0^T |x(s)|^2\, ds \right\}^{1/2}. \tag{11.21}$$

Then $\limsup |x|_T = |x|$ as $T \to \infty$, for any $x \in \mathcal{M}_2$. If we consider the truncated function $x_T(t)$, then Parseval's equality allows us to write

$$|x|_T = \left\{ (2\pi T)^{-1} \int_R |\tilde{x}_T(\omega)|^2\, d\omega \right\}^{1/2}. \tag{11.22}$$

We have also

$$|Ax|_T = \left\{ (2\pi T)^{-1} \int_R |\tilde{y}_T(\omega)|^2\, d\omega \right\}^{1/2}, \tag{11.23}$$

with

$$\tilde{y}_T(\omega) = \int_0^T e^{i\omega t}\left[ax(t) + \int_0^t k(t - \tau)x(\tau)\, d\tau \right] dt. \tag{11.24}$$

A simple calculation shows that (11.24) can be written in the form

$$\tilde{y}_T(\omega) = \tilde{x}_T(\omega)K(\omega) - \int_T^\infty e^{i\omega t}\left[\int_0^T k(t - \tau)x(\tau)\, d\tau \right] dt. \tag{11.25}$$

Let us denote the double integral occurring in the right member of (11.25) by $\eta(T, \omega)$. Then $\eta(T, \omega)$ is the Fourier transform of the function

$$Z_T(t) = \begin{cases} \int_0^T k(t - \tau)x(\tau)\,d\tau, & \text{for} \quad t > T, \\ 0, & \text{for} \quad t \leq T. \end{cases}$$

On the other hand, $Z_T(t)$ represents for $t > T$ the restriction of the convolution product $k * x_T$. Since $|k| \in L(R_+, R)$ and $|x_T| \in L^2(R_+, R)$ there results $|z_T| \in L^2(R_+, R)$. Parseval's equality leads to

$$T^{-1} \int_0^\infty |z_T(t)|^2 \, dt = (2\pi T)^{-1} \int_R |\eta(T, \omega)|^2 \, d\omega. \tag{11.26}$$

We shall now estimate the left side of (11.26). Denoting it by $\gamma^2(T)$, we get

$$\gamma^2(T) = \frac{1}{T} \int_T^{T+\sqrt{T}} dt \left| \int_0^T k(t - \tau)x(\tau)\,d\tau \right|^2 + T^{-1} \int_{T+\sqrt{T}}^\infty dt \left| \int_0^T k(t - \tau)x(\tau)\,d\tau \right|^2$$

$$= I_1 + I_2.$$

We have further

$$|I_1| \leq T^{-1}(|x|_{L^\infty})^2 \int_T^{T+\sqrt{T}} \left| \int_0^T |k(t - \tau)| \, d\tau \right|^2 dt$$

$$\leq T^{-1/2}(|x|_{L^\infty})^2 \left[\int_0^\infty |k(\tau)| \, d\tau \right]^2.$$

Consequently, $I_1 \to 0$ as $T \to \infty$. Since (11.16) implies $|k| \in L^2([T_0, \infty); R)$ for sufficiently large T_0, we can estimate I_2 as follows:

$$|I_2| \leq T^{-1} \int_{T+\sqrt{T}}^\infty \left[\int_0^T |k(t - \tau)|^2 \, d\tau \int_0^T |x(\tau)|^2 \, d\tau \right] dt$$

$$\leq (|x|_T)^2 \int_{T+\sqrt{T}}^\infty dt \int_{t-T}^t |k(\tau)|^2 \, d\tau$$

$$\leq M^2/2\varepsilon(1 + 2\varepsilon) + (|x|_T)^2/T^{-\varepsilon}.$$

Therefore, $I_2 \to 0$ as $T \to \infty$. From the preceding considerations one obtains

$$\lim_{T \to \infty} \gamma(T) = 0. \tag{11.27}$$

Taking into account (11.23), (11.25), (11.26), and (11.22), there results

$$|Ax|_T \leq A_0|x|_T + \gamma(T), \tag{11.28}$$

as long as T is sufficiently large. If we take the upper limit of both sides in (11.28) and consider (11.27), we obtain (11.20) for any $x \in E$.

Let us show that A_0 cannot be replaced by a smaller number with the same property. Indeed, if we take $x(t) = \exp(-i\omega_0 t)$, with $\omega_0 \in R$, then

$$(Ax)(t) = K(\omega_0)x(t) + \varepsilon(\omega_0, t), \qquad (11.29)$$

with

$$\varepsilon(\omega_0, t) = -\int_t^\infty k(\tau)e^{-i\omega(t-\tau)}\, d\tau.$$

Since

$$|\varepsilon(\omega_0, t)| \le \int_t^\infty |k(\tau)|\, d\tau \to 0$$

as $t \to \infty$, (11.29) yields $|Ax| \le |K(\omega_0)|\,|x|$. This proves that A_0 is the best constant in (11.20). Theorem 11.1 is completely proved.

We shall now prove a theorem on the energetic stability of Eq. (11.12), with A given by (11.13).

Theorem 11.2   Consider Eq. (11.12) and assume that the linear operator A is defined by (11.13), with $k(t)$ satisfying the conditions of Theorem 11.1. Moreover, let F be a (nonlinear) operator from L^∞ into itself such that

$$|F(x) + \lambda x| \le \gamma|x|, \qquad x \in L^\infty, \qquad (11.30)$$

where λ and $r > 0$ are certain constants. If the frequency condition

$$\inf_{\operatorname{Im} s \ge 0} |(1/K(s)) + \lambda| > r \qquad (11.31)$$

holds true, then Eq. (11.12) is energetically stable (in L^∞).

Proof   We shall reduce the proof of Theorem 11.2 to Lemma 11.2 and Theorem 11.1. The constant λ from (11.30) will be used with the meaning required by Lemma 11.2. It is obvious that (11.30) is equivalent to

$$\|F + \lambda I\| \le r.$$

Condition (11.9) will be satisfied if

$$\|(I + \lambda A)^{-1}A\| < r^{-1}. \qquad (11.32)$$

Now, we prove that (11.32) holds because of (11.31) and

$$\|(I + \lambda A)^{-1}A\| = \left[\inf_{\operatorname{Im} s \ge 0} |(1/K(s)) + \lambda|\right]^{-1}. \qquad (11.33)$$

There remains to show the validity of (11.33). We denoted the transfer function associated with A by $K(s)$. Then $I + \lambda A$ has $1 + \lambda K(s)$ as its transfer function. According to (11.31), there exists $(I + \lambda A)^{-1}$, and its transfer function will obviously be $[1 + \lambda K(s)]^{-1}$. Therefore, the transfer function corresponding to $(I + \lambda A)^{-1}A$ is $K(s)[1 + \lambda K(s)]^{-1}$. Consequently, Theorem 11.1 gives

$$\|(I + \lambda A)^{-1}A\| = \sup_{\text{Im } s \geq 0} |K(s)/[1 + \lambda K(s)]\| = \inf_{\text{Im } s \geq 0} |(1/K(s)) + \lambda|^{-1}.$$

Formula (11.33) is thus proved and this ends the proof of Theorem 11.2.

Remark   The reader not acquainted with system theory and fond of a pure mathematical argument in proving (11.33) could supply an alternative proof of this formula by using Theorem 1.3.6. The normed ring to which this theorem has to be applied consists of all functions that can be represented in the form $K(s) = a + \tilde{k}(s)$, where $\tilde{k}$ is the Laplace transform of a function k, with $|k| \in L(R_+, R)$.

Corollary   Consider Eq. (11.1) under the following assumptions:

1. the linear operator A given by (11.13) satisfies the conditions of Theorem 11.2;

2. $\varphi(t, \sigma)$ is a continuous function for $t \in R_+$, $\sigma \in R$, such that

$$m \leq \varphi(t, \sigma)/\sigma \leq M, \qquad \sigma \neq 0, \tag{11.34}$$

with m, M given numbers;

3. the frequency condition

$$\inf_{\text{Im } s \geq 0} |(1/K(s)) + (M + m)/2| > (M - m)/2 \tag{11.35}$$

holds true.

Then Eq. (11.1) is energetically stable (in L^∞). In other words, for any $h \in L^\infty$, with $|h| = 0$, the solution $\sigma(t)$ (whose existence in L^∞ is assumed!) of Eq. (11.1) also satisfies $|\sigma| = 0$.

The proof of this corollary is an immediate consequence of Theorem 11.2. Indeed, condition (11.34) implies condition (11.30) for the operator F given by (11.14), with $\lambda = \frac{1}{2}(M + m)$ and $r = \frac{1}{2}(M - m)$. It is then obvious that (11.31) becomes (11.35).

In concluding this section we should like to point out that in his paper [3], Beneš considered essentially the same kind of stability. Furthermore, his approach is based on a detailed discussion of the space $\mathcal{M}_2(R, R)$, which is considerably richer than $L^2(R, R)$ or $L^\infty(R, R)$. The existence of a solution for the integral equation

$$\sigma(t) = h(t) + \int_0^t k(t - \tau)\varphi(\tau, \sigma(\tau)) \, d\tau$$

is also proved by means of the contraction mapping principle.

3.12   A Criterion for L^p Stability

This section deals with the scalar integral equation of Volterra type

$$x(t) = h(t) + \int_0^t k(t - s)\varphi(s, x(s))\, ds, \qquad (12.1)$$

under conditions that assure the existence of an L^p solution for any h belonging to L^p, $1 \le p \le \infty$. The result is due to Grossman [1]. The interesting feature of the result we shall formulate below consists of the fact that the key conditions involved in the statement are independent of p, excepting those concerning the free term $k(t)$ and the nonlinearity $\varphi(t, x)$. This feature allows us to speak of L^p stability and to regard the result we are going to establish as a criterion of L^p stability.

The following assumptions are needed in order to state the main result of this section.

a. $\varphi(t, x)$ is a mapping from $R_+ \times R$ into R, continuous in x and measurable in t, such that there exist constants α and β, $\alpha < \beta$, $\beta > 0$, with the property

$$\alpha(x - y) \le \varphi(t, x) - \varphi(t, y) \le \beta(x - y) \qquad (12.2)$$

for any $x, y \in R$, $x \ge y$.

b.   There exists a function $\omega \in L(R_+, R)$ such that

$$m = k + k * \omega \in L(R_+, R),$$

where the star denotes the convolution product:

$$(k * \omega)(t) = \int_0^t k(t - s)\omega(s)\, ds. \qquad (12.3)$$

We define a function $\eta(s)$, $s \in R$, in order to measure the deviation from the average slope of φ:

$$\eta(s) = \begin{cases} s - \alpha, & s \ge \tfrac{1}{2}(\alpha + \beta), \\ \beta - s, & s < \tfrac{1}{2}(\alpha + \beta). \end{cases}$$

Let γ be a real number. It is obvious that Eq. (12.1) can be written equivalently as

$$x = h + \gamma(k * x) + k * [\varphi(x) - \gamma x], \qquad (12.4)$$

where $\varphi(x)(t) = \varphi(t, x(t))$. If $\omega(t)$ satisfies (b), then

$$w_\gamma(t) = \gamma m(t) - \omega(t) \qquad (12.5)$$

belongs to $L(R_+, R)$ for all constants γ. Hence, the Laplace transform $\tilde{w}_\gamma(s)$ is defined for all s with $\mathrm{Im}\, s \ge 0$.

Another hypothesis we shall need in this section is:

c. There exists at least one $\gamma \in R$ such that

$$\tilde{w}_\gamma(s) \neq 1 \qquad \text{for} \quad \text{Im } s \geq 0. \tag{12.6}$$

If γ satisfies (12.6), then there exists $n_\gamma(t) \in L(R_+, R)$ such that

$$[1 - \tilde{w}_\gamma(s)]^{-1} = 1 - \tilde{n}_\gamma(s), \qquad s \in R. \tag{12.7}$$

In other words, $n_\gamma(t)$ is the resolvent kernel for $w_\gamma(t)$. The existence of $n_\gamma(t)$ follows from Theorem 1.3.6.

The last condition we shall assume with respect to Eq. (12.1) can be stated as follows:

d. There exists at least one $\gamma \in R$ satisfying (12.6) and

$$\eta(\gamma)|m - m * n_\gamma|_L < 1. \tag{12.8}$$

Theorem 12.1 Let E be any of the spaces $L^p(R_+, R)$, $1 \leq p \leq \infty$. Assume that conditions (a)–(d) are verified for Eq. (12.1) and, furthermore, that $x \rightarrow \varphi(x)$ is a mapping from E into itself. Then there exists a unique solution for (12.1) lying in E for any $h \in E$.

Proof Let $\gamma \in R$ be a number satisfying conditions (c) and (d) of Theorem 12.1. Consider Eq. (12.4), which is equivalent to (12.1), for such a value of γ. We now define the function $v(t)$ by $v(t) = -\omega(t) - \int_0^t \omega(t - s)v(s) \, ds$, with $\omega(t)$ as described in condition (b) above. The existence and uniqueness of $v(t)$ in $L_{\text{loc}}(R_+, R)$ is obvious. The relationship between ω and v can be written in operational form as $(I + v)^{-1} = I + \omega$. If we add $v * x$ to both sides of Eq. (12.4), there results

$$x + v * x = h + \gamma(k + v) * x + k * [\varphi(x) - \gamma x],$$

which yields

$$x = (I + v)^{-1} * h + (I + v)^{-1} * (\gamma k + v) * x + (I + v)^{-1} * k * [\varphi(x) - \gamma x]. \tag{12.9}$$

If we denote $f(t) = (I + v)^{-1} * h(t)$, we get $f \in E$, because $f = h + \omega * h$ and $\omega \in L(R_+, R)$. We have further $(I + v)^{-1} * (\gamma k + v) = (I + \omega) * (\gamma k + v) = \gamma k + \gamma \omega * k + v + \omega * v = \gamma(k + \omega * k) - \omega = w_\gamma$. Similarly, $(I + v)^{-1} * k = (I + \omega) * k = k + \omega * k = m$. Consequently, (12.9) becomes

$$x = f + w_\gamma * x + m * [\varphi(x) - \gamma x]. \tag{12.10}$$

Transforming further we get from (12.10)

$$x - w_\gamma * x = f + m * [\varphi(x) - \gamma x]$$

and
$$x = (I - w_\gamma)^{-1} * f + (I - w_\gamma)^{-1} * m * [\varphi(x) - \gamma x].$$
But $(I - w_\gamma)^{-1} = I - n_\gamma$. Let $g = (I - w_\gamma)^{-1} * f = f - n_\gamma * f \in E$. Hence
$$x = g + (m - m * n_\gamma) * [\varphi(x) - \gamma x]. \tag{12.11}$$

This equation is equivalent to (12.1) and is particularly adequate for treatment by means of the contraction mapping principle. If we denote by Tx the right-hand side of (12.11), then $x \to Tx$ is obviously a mapping from E into itself.

Moreover,
$$
\begin{aligned}
|Tx - Ty|_E &\le |m - m * n_\gamma|_L \, |\varphi(x) - \varphi(y) - \gamma(x - y)|_E \\
&\le |m - m * n_\gamma|_L \, |[(\varphi(x) - \varphi(y))/(x - y) - \gamma](x - y)|_E \\
&\le |m - m * n_\gamma|_L \, \eta(\gamma) |x - y|_E .
\end{aligned}
$$

Taking into account condition (12.8), there results that T is a contraction mapping. This ends the proof of Theorem 12.1.

Remark 1 If we assume that $\varphi(t, x)$ is continuous in (t, x), then the statement of Theorem 12.1 is also valid for $E = C(R_+, R)$.

Remark 2 Condition (b) is verified for any $\omega \in L(R_+, R)$ if $k \in L(R_+, R)$. It is interesting to know whether this condition holds for some $k \notin L(R_+, R)$. For example, if $k(t) = -1 + e^{-t} \notin L(R_+, R)$ and $\omega(t) = -e^{-t}$, then $m(t) = k(t) + (k * \omega)(t) = -te^{-t} \in L(R_+, R)$. In other words, condition (b) can be satisfied for nonintegrable kernels on R_+.

Remark 3 It follows easily from condition (12.2) that
$$|\varphi(t, x) - \varphi(t, y)| \le L|x - y|, \tag{12.12}$$
where $L = \max\{|\alpha|, \beta\}$. If we assume that $\varphi(t, 0) \in L^p$ for a certain p,
$$1 \le p \le \infty,$$
then (12.12) yields $\varphi(t, x(t)) \in L^p$ for any $x(t) \in L^p$. In particular, for $\varphi(t, 0) \equiv 0$ —a condition that is usually encountered—we see that $x \to \varphi(x)$ is a mapping from L^p into itself, for any p with $1 \le p \le \infty$.

Exercises

1. Consider the integro-differential system

(S) $\qquad \dot{x}(t) = Ax(t) + \displaystyle\int_0^t b(t - s)\varphi(\sigma(s))\, ds, \qquad \sigma = (c, x),$

and find an integral equation for $\sigma(t)$. If A is a Hurwitzian matrix and $b(t) \in L(R_+, R)$, then a sufficient condition for the absolute stability of (S) with respect to the class of all continuous $\varphi(\sigma)$, with $\sigma\varphi(\sigma) > 0$ for $\sigma \neq 0$, is: there exists $q > 0$ such that

$$\text{Re}\{(isq - 1)(c, (isI + A)^{-1}\tilde{b}(s))\} \leq 0$$

for any $s \in R$.

2. Investigate the absolute stability of the system

$$\dot{x}(t) = Ax(t) + \int_0^t b(t - s)\varphi(\sigma(s))\, ds, \qquad \dot{\xi} = \varphi(\sigma), \quad \sigma = (c, x).$$

Hint: Find an integral equation for $\sigma(t)$ and apply Theorem 3.1.

3. Investigate the existence and stability properties of the solutions of the following integro-differential system:

$$\dot{x} = Ax + b\varphi(\sigma), \qquad \sigma(t) = f(t) + \int_0^t (c(t - s), x(s))\, ds.$$

Show that the frequency condition can be written as

$$\text{Re}\{(isq - 1)(\tilde{c}(s), (isI + A)^{-1}b)\} \leq 0,$$

where $q \geq 0$ is a certain number and $s \in R$.

4. Consider the integral equation

(E) $$\sigma(t) = h(t) + \int_0^t k(t - s)\varphi(\sigma(s))\, ds,$$

where $\sigma, h, \varphi \in R^n$, and $k(t)$ is a matrix kernel of type n by n. Assume that the following conditions are verified: (a) $h'(t)$, $h''(t) \in L(R_+, R^n)$; (b) $k(t) = k_0(t) - \Gamma$, with Γ a positive definite constant matrix and $\|k_0(t)\|$, $\|k_0'(t)\| \in L(R_+, R)$; (c) $\varphi(\sigma)$ is continuous on R^n, $(\sigma, \varphi(\sigma)) \geq 0$ for $\sigma \neq 0$, and there exist a matrix Q and a positive definite function $F(\sigma)$ such that $dF = (\varphi(\sigma), Q\, d\sigma)$; (d) for any $s \in R$, we have

$$\text{Re}\{(I - isQ)[\tilde{k}_0(s) + \Gamma(is)^{-1}]\} \leq 0.$$

Then any solution $\sigma(t)$ of Eq. (E) belongs to $C_0(R_+, R^n)$.

5. Discuss the frequency conditions of Theorems 2.2 and 3.1 when $\varphi(\sigma)$ is subject to the following restriction: $0 < \sigma\varphi(\sigma) \leq \beta\sigma^2$, with $\beta > 0$.

6. Discuss the existence problem of solutions for Eq. (E) from Exercise 4 above, under assumptions of Theorem 6.1.

7. Consider the equation

(E) $$x(t) = h(t) + \int_0^\infty k(t - s)\varphi(s; x_s)\, ds$$

under following conditions: (a) $h(t) \in C(R_+, R^n)$, i.e., $h(t)$ is a continuous bounded function from R_+ into R^n; (b) $\|k(t)\| \in L(R, R)$; (c) if x_s denotes the restriction of $x(t)$ to $[0, s]$, then the mapping $x \to \varphi(t; x_t)$ is continuous from $C_c(R_+ R^n)$ into itself and, moreover, there exists $M > 0$ such that

$$\|\varphi(t; x_t)\| \le M$$

for any $x \in C_c(R_+, R^n)$ and $t \in R_+$. Then there exists at least one solution of (E) belonging to $C(R_+, R^n)$.

8. Consider the equation

(E) $$x(t) = h(t) + \int_0^t k(t - s)\varphi(s; x_s) \, ds$$

under following conditions: (a) $h(t)$, $h'(t) \in L(R_+, R)$, and $h'(t) \in L^2_{\text{loc}}(R_+, R)$; (b) $|k(t)|$, $|k'(t)| \in L(R_+, R)$; (c) the same as in Exercise 7, with $n = 1$; (d) for any $x(t) \in C(R_+, R)$, $\varphi(t; x_t)x(t) \ge 0$ on R_+, and for any $\varepsilon > 0$, there exists $\delta > 0$ such that $(|x(t)| > \varepsilon) \Rightarrow (\varphi(t; x_t)x(t) > \delta)$; (e) for every $x(t) \in C(R_+, R)$ with $x'(t) \in L_{\text{loc}}(R_+, R)$, there exists a positive constant λ such that

$$\int_0^t \varphi(s; x_s) \, dx(s) \ge -\lambda, \qquad t \in R_+;$$

(f) there exists $q \ge 0$ such that

$$\text{Re}\{(1 - isq)\tilde{k}(s)\} \le 0, \qquad s \in R.$$

Then there exists at least one solution of (E) belonging to $C_0(R_+, R)$.

9. Let us consider the system

(S) $$\dot{x} = Ax + bf(\sigma), \qquad \sigma = -(c, x) - hf(\sigma),$$

where x denotes a real n-vector, A is a real $n \times n$ matrix whose eigenvalues may include zero, b and c are real n-vectors, and h is a nonnegative constant. The nonlinear function $f(\sigma)$ is continuous for $\sigma \in R$, satisfies the inequalities $0 < \sigma f(\sigma) < k\sigma^2$, $\sigma \ne 0$, and there exists $f'(\sigma)$, which is also continuous and such that $1 + hf'(\sigma) \ne 0$, $\sigma \in R$. Show that any solution of (S) approaches zero as $t \to \infty$, provided the frequency condition

$$k^{-1} + \text{Re}\{(isq - 1)(c, (isI + A)^{-1}b)\} + h \ge 0$$

is verified for any $s \in R$.

10. Let us consider the nonlinear time-invariant feedback system described as follows: The linear time-invariant subsystem is characterized by the input–output equation

(L) $$y(t) = z(t) + \int_0^t g(t - s)\alpha(s) \, ds, \qquad t \in R_+,$$

where $g(t) \in L(R_+, R)$, and the zero-input response $z(t)$ is such that

$$z(t) \in C_0(R_+, R) \cap L^2(R_+, R)$$

and $\dot{z}(t) \in L^2(R_+, R)$. The nonlinear time-invariant memoryless subsystem is given by the equation (N): $\alpha(t) = \phi(\eta(t))$, with $\phi(\eta)$ continuous on R and satisfying the inequalities $0 < \varepsilon \le \phi(\eta)/\eta \le k - \varepsilon$ for all $\eta \ne 0$. The single-loop feedback system characterized by (L), (N), and $\eta = -y$ is stable [i.e., $y(t) \in C_0(R_+, R)$], if there exists $q \ge 0$ and a positive δ such that

$$\mathrm{Re}\{(1 - isq)\tilde{g}(s) + k^{-1}\} \ge \delta.$$

Hint: Apply the procedure used in Sections 3.2 and 3.3 to the integral equation $-\eta(t) = z(t) + \int_0^t g(t - s)\phi(\eta(s))\, ds$.

11. Consider the single-loop feedback system governed by the differential equations $q(D)x = y$, $p(D)x + f(y) = 0$, under the following assumptions: (a) $p(x)$ and $q(x)$ are polynomials without common factors, and the degree of $p(x)$ exceeds that of $q(x)$; (b) $f(y)$ is a continuous function for $y \in R$ such that $0 < yf(y) < ky^2$, for any $y \ne 0$; (c) there exists a real number α such that $(1 + \alpha s)G(s) + k^{-1}$, with $G(s) = q(s)/p(s)$, is positive real. Then the feedback system under consideration is asymptotically stable in the large (i.e., any solution approaches zero as $t \to \infty$).

Remark We recall that a rational function $Z(s)$ is *positive real* if the following conditions hold: (i) $Z(s)$ takes real values for real s; (ii) $\mathrm{Re}\, Z(i\omega) > 0$ for all real ω; (iii) $Z(s)$ has no right half-plane poles; (iv) all the imaginary axis poles of $Z(s)$ are simple and the corresponding residues are positive.

12. Consider the integral equation

(E) $$\sigma(t) = h(t) + \lambda_0 + \mu_0 t - \int_0^t [k(t - s) + \alpha + \beta(t - s)]\varphi(\sigma(s))\, ds$$

under the following assumptions: (a) $h(t) \in L(R_+, R) \cap L^2(R_+, R)$, $h'(t) \in L(R_+, R)$; (b) $k(t)$, $k'(t) \in L(R_+, R)$; (c) λ_0, μ_0, α, and β are real and $\alpha > 0$, $\beta > 0$; (d) $\varphi(\sigma)$ is continuous on R, $0 < \sigma\varphi(\sigma)$ for $\sigma \ne 0$ and

$$\lim_{|\sigma| \to \infty} \int_0^\sigma \varphi(u)\, du = +\infty;$$

(e) $\alpha - s\, \mathrm{Im}\, \tilde{k}(s) > 0$ for $s \ge 0$ and $\lim_{s \to \infty} [\alpha - s\, \mathrm{Im}\, \tilde{k}(s)] > 0$. Then $\sigma(t) \in C_0(R_+; R)$, where $\sigma(t)$ is the solution of Eq. (E).

13. Consider the vector integral equation

(E) $$\sigma(t) = h(t) + \int_0^t k(t - s)f(s, \sigma(s))\, ds$$

and assume that: (a) $h(t) \in L^2(R_+, R^n)$; (b) $f(t, \sigma)$ is continuous in σ, $\sigma \in R^n$, measurable in t for fixed σ, and such that $\|f(t, \sigma)\| \leq L\|\sigma\| + c(t)$, with $L > 0$ and $c(t) \in L^2(R_+, R)$; (c) $k(t)$ is a square matrix kernel of type n by n and $\|k(t)\| \in L(R_+, R) \cap L^2(R_+, R)$; (d) there exists a real number q, $|q| < L^{-1}$, such that the Hermitian matrix $q(\tilde{k} + \tilde{k}^*) - 2\tilde{k}\tilde{k}^*$ is positive semidefinite. Then there exists at least one solution $\sigma(t)$ of Eq. (E) such that $\sigma(t) \in L^2(R_+, R^n)$.

   14. Investigate the absolute stability of the feedback system described by the following equations with delay ($h > 0$):

$$\dot{x}(t) = Ax(t) + Bx(t - h) + b\xi(t) + c\xi(t - h),$$

$$\ddot{\xi}(t) + \alpha\dot{\xi}(t) + \beta\xi(t) = f(\sigma(t)),$$

$$\sigma = (p, x) + \gamma\xi + \delta\dot{\xi}.$$

Assume that the characteristic equation $\det(A + Be^{-\lambda h} - \lambda I) = 0$ has its zeros in the half-plane Re $\lambda < 0$ and $0 < \sigma f(\sigma) \leq k\sigma^2$, $\sigma \neq 0$, $\int_0^{\pm\infty} f(\sigma)\,d\sigma = +\infty$. Find the frequency condition.

Bibliographical Notes

   The results of Sections 3.2 and 3.3 are due to the author [1, 2]. Section 3.4 closely follows a paper by Lellouche [1]. The result of Section 3.5 belongs to Geleg [2], who also considered nonintegrable kernels of the form shown in Exercise 12 above, and even kernels of a more sophisticated nature. The stability result given in Section 3.6 is due to Doležal [1], and that included in Section 3.7 was obtained by Moser [1]. The problem of existence of L^2 solutions treated in Section 3.8 seems to be the first nonclassical stability result obtained by using Popov's technique. The results of Section 3.10 are due to Yakubovitch [1], those of Section 3.11 on energetic stability belong to Kudrewicz [2], and the result in Section 3.12 was recently obtained by Grossman [1].

   The exercises pertaining to this chapter were selected from the author's papers [4, 7], Rao's papers [1, 2], Luca's paper [1], Halanay's papers [1, 3], and from the results obtained by Ku and Chieh [1], Brockett and Willems [1], and Desoer [1].

   The first paper by Popov in which the frequency technique combined with the truncation procedure is used in studying the stability of nonlinear feedback systems appeared in 1959 [1]. It is interesting to point out that the system investigated by Popov in his paper [1] is governed by an integral equation. Papers [3] and [4] of Popov are devoted to the development of his method, but only systems governed by differential equations are considered. In his

book "Hyperstability of Automatic Control Systems" [5], Popov thoroughly
investigates various classes of systems. Among them, there are some systems
described by integral operators. Absolute stability is considered as a particular
case of the concept of hyperstability.

Since 1961—when Popov's method became generally known—a consider-
able number of research papers devoted to the investigation of stability of
various classes of nonlinear control systems have been written. Popov's
method and its generalizations were extensively used in applied mathematics
and engineering, and they provide the most salient and effective criteria of
stability and asymptotic behavior for various classes of control systems. A
common feature of the systems to which Popov's techniques can be successfully applied consists of the fact that they contain a linear time-invariant subsystem whose unit impulse response is integrable (in other words, the kernel
belongs to the space L).

We shall now review some of the most relevant achievements related to
the use of frequency techniques in the study of feedback systems. Of course,
the main attention is paid to those results that are related to the content of
this chapter.

Results that are similar to those given in Sections 3.2–3.5 were obtained
by Albertoni *et al.* [1], Albertoni and Szegö [1], Szegö [1], Blodgett and King
[1], [2], and Dewey [1]. Many of these results regard feedback systems with
various restrictions on the nonlinearity (e.g., boundedness, slope restrictions,
sector restrictions). The unit impulse response of the linear subsystem is assumed integrable or of the form considered in Section 3.3. The results of
Sections 3.2 and 3.3 can be found in the book by Reissig *et al.* [1].

The concept of energetic stability was discussed and extended in the following papers: Kudrewicz [3], Beneš [3], and Doležal [4]. The paper by
Beneš is particularly recommended because it contains a full description of
the space $\mathcal{M}_2$.

Let us remark that the frequency condition (11.31) encountered in
establishing the result of Kudrewicz can be easily interpreted as follows: the
graph of the function $z = 1/K(s), s \in R$, is situated outside the circle $|z + \lambda| = r$.
Very often, such a frequency condition is called a "circle criterion." Similar
frequency conditions were used by Sandberg [2] and Zames [1] in connection
with stability theory of some classes of dynamical systems.

Among the major contributors to the theory of stability of nonlinear
feedback systems we should like to mention Yakubovitch. His papers are
mainly concerned with systems described by ordinary differential equations
and did not find place in our list of references. Some interesting results due
to this author can be found in the book by Aizerman and Gantmacher [1],
in that by Lefschetz [1], and in various papers he published in *Automation and
Remote Control, Doklady Akademii Nauk SSSR,* and *Vestnik Leningradskogo*

Universiteta. He uses the theory of matrix inequalities and the Liapunov's function method.

Recently, Taylor and Narendra [1] investigated the stability properties of a class of nonlinear time-varying systems of the form $\dot{x} = Ax - bk(t)f(\sigma)$, $\sigma = (c, x)$, $k(t) \geq 0$, using simultaneously the frequency techniques and the Liapunov's function method. J. L. Willems [1] considers the system $\dot{x} = Ax - b[f(t, y) - r(t)]$, $y = (c, x)$, $f(t, 0) \equiv 0$, and studies the following kind of stability: the system under consideration is called stable if $r(t) \in L^2(R_+, R)$ implies $y(t) \in L^2(R_+, R^n) \cap C_0(R_+, R^n)$. The main achievement of this paper consists of the fact that the maximum generality is allowed for the nonlinear function $f(t, y)$. By adequately choosing some constants used in defining the class of nonlinearities, one obtains practically all the situations encountered in the literature.

The stability properties of solutions of Volterra integral equations have been investigated in a series of papers by Sandberg [1–5]. His paper [4] surveys the main results he obtained by systematically using the frequency techniques and the truncation procedure. A typical result of Sandberg [1, 4] can be formulated as follows: Consider the integral equation

(E) $$f(t) + \int_0^t k(t - s)\psi(s, f(s)) \, ds = g(t), \qquad t \in R_+,$$

under the following conditions: (a) $g(t) \in L^2(R_+, R^n)$ is given; (b) $k(t)$ is a matrix kernel of type n by n such that $\|k(t)\| \in L(R_+, R)$; (c) there exist two real numbers α and β, with $\alpha \leq \beta$, such that (i) $\det[I + \frac{1}{2}(\alpha + \beta)\tilde{k}(s)] \neq 0$ for $\operatorname{Im} s \geq 0$ and (ii) $\sup_{s \in R} \Lambda\{[I + \frac{1}{2}(\alpha + \beta)\tilde{k}(s)]^{-1}\tilde{k}(s)\} \leq 2(\beta - \alpha)^{-1}$ where $\Lambda(M)$ denotes the positive square root of the largest eigenvalue of the matrix M^*M; (d) $\psi(t, w)$ is a mapping from $R_+ \times R^n$ into R^n, continuous in w for fixed t and measurable in t for fixed w, such that $\alpha \leq \psi_h(t, w)/w \leq \beta$, $w \neq 0$, $h = 1, 2, \ldots, n$. Then any solution $f(t) \in L^2_{\text{loc}}(R_+, R^n)$ of Eq. (E) belongs to $L^2(R_+, R^n)$.

The linear subsystems of the feedback systems described by the integral equations investigated in this chapter are characterized by an input–output equation of the form $y(t) = h(t) + \int_0^t k(t - s)x(s) \, ds$, $t \in R_+$. A single exception constitutes the linear subsystem associated with Eq. (11.1). The input–output equation has the form (11.13)—under assumption that the zero-input response is zero. Since (11.13) can be also written as

$$y(t) = (Ax)(t) = ((k + \delta) * x)(t),$$

where δ denotes the Dirac function, it is clear that in this way the impulses may be included in the linear subsystem. An important number of papers are devoted to the investigation of feedback systems whose linear subsystems are described by input–output equations of the form

$$y(t) = \sum_{n=1}^{\infty} g_n x(t - t_n) + \int_0^{\infty} g(t - s)x(s)\, ds, \qquad t \in R_+ .$$

In the above equation, $\{t_n\}$ represents a sequence of nonnegative real numbers, $\{g_n\}$ is a sequence of real numbers with $\sum |g_n| < +\infty$, and $y(t) \in L(R, R)$. It can be assumed, of course, that $\{g_n\}$ is a sequence of m by m matrices and $y(t)$ is also a matrix function of the same type. The unit impulse response is then given by $g_1(t) = g(t) + \sum_{n=1}^{\infty} g_n \delta(t - t_n)$. The use of distributions appears to be very convenient in order to describe systems with impulses. On these lines, we mention the papers of Zames and Falb [1, 2], Desoer and Wu [1, 2, 3], Wu and Desoer [1], J. C. Willems [2], and Barbălat and Halanay [1]. A more general case was recently investigated by Baker and Vakharia [1]. The unit impulse response is of the form

$$g_2(t) = g_1(t) + \sum_{k=0}^{m} a_k t^k + \sum_{k=0}^{p} b_k t^{\alpha_k} e^{\lambda_k t},$$

with $g_1(t)$ as above.

We shall now give, for illustration, a result due to Wu and Desoer [1]. This result provides a criterion of L^p stability for a feedback nonlinear system governed by the equations $y_1(t) = (g_1 * e_1)(t)$, $y_2 = \varphi(e_2(t), t)$, $e_1 = u_1 - y_2$, and $e_2 = u_2 + y_1$. The unit impulse response $g_1(t)$ is of the form

$$r + \sum_{n=1}^{\infty} g_n \delta(t - t_n) + g(t),$$

with $r \geq 0$, $g(t) \in L(R_+, R)$, and $\sum_{n=1}^{\infty} |g_n| < +\infty$. The nonlinearity φ is subject to the following restrictions: $\varphi(\sigma, t)$ is a continuous mapping from $R \times R_+$ into R and for some constants k_1, k_2 we have $k_1 \sigma^2 \leq \sigma\varphi(\sigma, t) \leq k_2 \sigma^2$ for all $t \in R_+$ and $\sigma \in R$. Assume further that there exists a constant $k \in [k_1, k_2]$ with $kr > 0$ and such that $\inf |1 + k\tilde{g}(s)| > 0$, where the inf is taken with respect to all s in the half-plane Im $s \geq 0$. Moreover, if $\lambda = \max\{|k_2 - k|, |k_1 - k|\}$, then $\|h_k\|\lambda < 1$, where $h_k(t) = \sum_{n=1}^{\infty} \ell_n \delta(t - t_n) + h(t)$ is such that its Laplace transform equals $[1 + k\tilde{g}(s)]^{-1}\tilde{g}(s)$, and $\|h_k\| = \sum_{n=1}^{\infty} |\ell_n| + \int_0^{\infty} |h(t)|\, dt$. Then for any inputs u_1, $u_2 \in L^p(R_+, R)$ such that the corresponding e_1, $e_2 \in L^2_{loc}(R_+, R)$, there results e_1, e_2, y_1, $y_2 \in L^p(R_+, R)$. The above statement holds true for any p, with $1 \leq p \leq \infty$. The proof of this L^p stability result lies upon the fact that the distributions which can be represented in the form $\sum_{n=1}^{\infty} g_n \delta(t - t_n) + g(t)$, with $\sum_{n=1}^{\infty} |g_n| < \infty$ and

$$g(t) \in L(R_+, R),$$

can be organized as a Banach algebra (commutative normed ring).

Another interesting result we want to state here was obtained by Barbălat and Halanay [1]. They consider linear systems described by the input–output equation

(E) $$y(t) = \sum_{n=1}^{\infty} g_n x(t - t_n) + \int_{-\infty}^{t} k(t - s)x(s) \, ds,$$

where $\{g_n\}$ is a sequence of complex numbers with $\sum_{n=1}^{\infty} |g_n| < \infty$, and $k(t)$ is a mapping from R_+ into the complex number field such that

$$|k(t)| \in L(R_+, R) \cap L^2(R_+, R).$$

It is assumed that $\{t_n\}$ is a sequence of nonnegative numbers. Let us associate with (E) the function

$$\eta(-\infty, t) = \text{Re} \int_{-\infty}^{t} \bar{x}(s)y(s) \, ds,$$

where $\bar{x}$ denotes, as usual, the conjugate of x. The input $x(t)$ is supposed to be a piecewise continuous function on R, with complex values such that

$$|x(t)| \leq M \exp(\alpha t)$$

for $t \leq 0$, where M and α are positive numbers (depending on x). By definition, the system described by (E) is *hyperstable* if $\eta(-\infty, t) \geq 0$ for any $t \in R_+$. Let us consider now the *transfer function*

$$\gamma(s) = \sum_{n=1}^{\infty} g_n e^{-st_n} + \int_{0}^{\infty} e^{-st} k(t) \, dt.$$

Then, a necessary and sufficient condition that the linear system under consideration by hyperstable is Re $\gamma(i\omega) \geq 0$ for $\omega \in R$. An application of the above result is made to the theory of nuclear reactors.

It is interesting to notice that many authors have used the methods of functional analysis in order to derive frequency criteria of stability for various classes of nonlinear systems. We already mentioned that the theory of Banach algebras provides a very useful tool in this respect. Several other concepts and methods of functional analysis are of current use in studying feedback systems. Besides the papers by Desoer, Wu and Desoer, and Sandberg we have quoted above, we should like to mention here the papers by Kudrewicz [1], Zames [1, 2], J. C. Willems [1] and Freedman *et al.* [1]. Interesting comments concerning the extent of functional analysis methods in control theory have been made by J. C. Willems [1]. The paper by Freedman *et al.* [1] contains a very general theory, applicable to various classes of nonlinear feedback systems, among them being those described by integral equations of the form

$$\sigma(x, t) = h(x, t) + \int_{0}^{t} \int_{0}^{1} k(x, y, t - \tau)\varphi(\sigma(y, \tau)) \, dy \, d\tau.$$

The case of systems described by some partial differential equations is also investigated.

In concluding these notes, we shall make one more remark with respect to the frequency conditions encountered in the statements of various results of Chapter 3. All these conditions are expressing the positivity of a certain function or of a quadratic form or, simply, the fact that a certain number conveniently associated to the system is positive. As shown in Popov's book [5], in Halanay's paper [5], and in the papers by Kudrewicz [3] and Zames [2], such conditions are related to the positivity of certain operators occurring in the description of the system. An example of this kind will be treated in the last chapter of this book.

4

Wiener-Hopf Equations

The main goal of this chapter is the study of Wiener–Hopf equations of the form

$$x(t) = h(t) + \int_0^\infty k(t - s)x(s)\, ds, \qquad t \in R_+, \qquad \text{(WH)}$$

under assumptions that allow the use of the Fourier transform. The theory we shall develop below is due, essentially, to M. G. Krein [1]. Its origin can be traced, of course, to Wiener's work concerning harmonic analysis and related topics (around 1930). But it found a definitive place in classical analysis in the years following the publication of M. G. Krein's paper [1]. The idea of considering the space M as underlying space seems to have been used first by the author. Any space used in Krein's theory is a subspace of M, and this property gives the possibility of a unified treatment of some aspects encountered in discussing Eq. (WH).

We shall also be concerned with some nonlinear problems for integral equations associated with (WH). Such equations are usually obtained by perturbing the linear equation (WH).

4.1   A Special Class of Kernels

Under suitable assumptions regarding the kernel $k(t)$ of Eq. (WH), it is possible to find a resolvent kernel, say $\gamma(t, s)$, such that the solution of (WH) is given by

$$x(t) = h(t) + \int_0^\infty \gamma(t, s)h(s)\, ds, \qquad t \in R_+ . \tag{1.1}$$

The integral operator Γ defined by

$$(\Gamma h)(t) = \int_0^\infty \gamma(t, s)h(s)\, ds, \qquad t \in R_+ , \tag{1.2}$$

plays a central role in studying Eq. (WH). We shall now give the main properties of the operator Γ, under assumptions for $\gamma(t, s)$ that—as shown in the next section—are automatically verified by the resolvent kernel we shall construct.

The following assumptions will be made with respect to $\gamma(t, s)$:

a. $\gamma(t, s)$ is a measurable complex-valued function on $R_+ \times R_+$ such that there exists $k_0(t) \in L(R, R)$ with the property

$$|\gamma(t, s)| \le k_0(t - s), \qquad 0 \le t, \ \ s < +\infty; \tag{1.3}$$

b. the mapping $t \to \gamma(t, \cdot)$ from R_+ into $L(R_+, \mathscr{C})$ is uniformly continuous, i.e., for any $\varepsilon > 0$, there exists $\delta = \delta(\varepsilon) > 0$ such that

$$\int_0^\infty |\gamma(t + \tau, s) - \gamma(t, s)|\, ds < \varepsilon \qquad \text{for} \quad |\tau| < \delta; \tag{1.4}$$

c. we have

$$\int_0^\infty \gamma(t, s)\, ds \in C_\ell(R_+, \mathscr{C}). \tag{1.5}$$

Let us remark that assumption (a) implies that $t \to \gamma(t, \cdot)$ is a mapping from R_+ into $L(R_+, \mathscr{C})$. From assumptions (a) and (b) there results that the function occurring in (1.5) is a continuous function (even uniformly continuous) on R_+.

We are now able to establish the basic property of the operator Γ.

Theorem 1.1 Let us consider the integral operator Γ, formally given by (1.2). If $\gamma(t, s)$ satisfies the conditions (a), (b), and (c), then Γ is a continuous operator from the space E into itself, where E stands for any of the spaces $M(R_+, \mathscr{C})$, $L^p(R_+, \mathscr{C})$ $(1 \le p \le \infty)$, $C(R_+, \mathscr{C})$, $C_\ell(R_+, \mathscr{C})$, or $C_0(R_+, \mathscr{C})$. The norm of Γ satisfies

$$\|\Gamma\|_E \le |k_0|_L. \tag{1.6}$$

Proof   Using only condition (a), we easily get $\Gamma M \subset M$, $\Gamma L^p \subset L^p$
$(1 \le p \le \infty)$, if we consider Theorem 7.1. From $\Gamma L^\infty \subset L^\infty$ and condition (b),
there results $\Gamma L^\infty \subset C$. Moreover, any function $(\Gamma h)(t)$, with $h \in L^\infty$, is uniformly continuous on R_+. It is then obvious that $\Gamma C \subset C$.

Inasmuch as any $h \in C_\ell(R_+, \mathscr{C})$ can be represented as $h(t) = h(\infty)$
$+ [h(t) - h(\infty)]$, and $[h(t) - h(\infty)] \in C_0(R_+, \mathscr{C})$, it suffices to discuss the case
$E = C_0$. For $h \in C_0$, we have

$$\left| \int_0^\infty \gamma(t, s)h(s)\, ds \right| \le \int_0^T k_0(t - s)|h(s)|\, ds + \int_T^\infty k_0(t - s)|h(s)|\, ds$$

$$\le |h|_C \int_{t-T}^t k_0(u)\, du + |k_0|_L \sup_{t \ge T} |h(t)|.$$

If $\varepsilon > 0$ is given, then we can choose $T = T(\varepsilon)$ sufficiently large such that

$$\sup_{t \ge T} |h(t)| < \varepsilon/2|k_0|_L.$$

With T fixed as above and $t \ge T_1(\varepsilon)$, we have

$$\int_{t-T}^t k_0(u)\, du < \varepsilon/2|h|_C.$$

From the above inequalities there results

$$\left| \int_0^\infty \gamma(t, s)h(s)\, ds \right| < \varepsilon \qquad \text{for} \quad t \ge T_1(\varepsilon),$$

which shows that $\Gamma h \in C_0$ for $h \in C_0$. Therefore, the statement of the theorem
holds for $E = C_0$.

There remains to consider the case $E = C_\ell(R_+, \mathscr{C})$. Condition (c) is
needed in order to ensure the inclusion $\Gamma C_\ell \subset C_\ell$. From condition (c) we see
that $\Gamma h \subset C_\ell$ for any $h \equiv$ constant. Taking into account the above representation for $h \in C_\ell$ and the fact $\Gamma C_0 \subset C_0$, we easily get $\Gamma h \in C_\ell$ for any $h \in C_\ell$.

The estimate (1.6) for $\|\Gamma\|_E$ also follows from Theorem 7.1. Once estab-
lished for $E = L^\infty$, it holds for $E = C, C_\ell, C_0$. Theorem 1.1 is completely
proven.

Remark 1   A trivial example of a kernel satisfying conditions (a), (b), and
(c) is given by $\gamma(t, s) = k(t - s)$, with $k \in L(R, C)$. Then $k_0(t) = |k(t)|$.

Remark 2   Another example of a kernel satisfying assumptions (a), (b),
and (c) is given by

$$\gamma(t, s) = k_1(t - s) + k_2(s - t) + \int_0^\infty k_1(t - u)k_2(s - u)\, du, \qquad (1.7)$$

with $k_1, k_2 \in L(R, \mathscr{C})$ such that

$$k_1(t) = k_2(t) = 0 \qquad \text{for} \quad t < 0. \qquad (1.8)$$

Indeed, we can take

$$k_0(t) = |k_1(t)| + |k_2(-t)| + \int_R |k_1(t-u)k_2(-u)| \, du. \tag{1.9}$$

Let us remark that the integral occurring in (1.9) is the convolution product of the functions $|k_1(t)|$ and $|k_2(-t)|$, both from $L(R, R)$. This shows that $k_0 \in L(R, R)$. To prove property (b), we observe that

$$\int_0^\infty |\gamma(t+\tau, s) - \gamma(t, s)| \, ds \le \int_R |k_1(t+\tau) - k_1(t)| \, dt$$

$$+ \int_R |k_2(t+\tau) - k_2(t)| \, dt + \int_R |k_1(t+\tau) - k_1(t)| \, dt \int_0^\infty |k_2(t)| \, dt.$$

Finally, simple transformations allow us to write

$$\int_0^\infty \gamma(t, s) \, ds = \int_0^t k_1(s) \, ds + \int_0^\infty k_2(s) \, ds + \int_0^\infty \int_0^\infty k_1(t-u)k_2(s-u) \, du$$

$$= \int_0^t k_1(s) \, ds + \int_t^\infty k_2(s) \, dt + \int_0^t k_1(u) \, du \int_0^\infty k_2(s) \, ds.$$

Now it is clear that $\gamma(t, s)$ given by (1.7) also satisfies condition (c).

The procedure we described above for constructing kernels $\gamma(t, s)$ will be used in the next section.

4.2   Equation (WH) with Index Zero

The main assumptions under which Eq. (WH) will be investigated in this section are:

$$k(t) \in L(R, \mathscr{C}), \tag{2.1}$$

$$1 - \tilde{k}(s) \ne 0, \qquad s \in R, \tag{2.2}$$

where $\tilde{k}(s)$ denotes the Fourier transform of $k(t)$, and

$$v = -\mathrm{ind}[1 - \tilde{k}(s)] = -(2\pi)^{-1}[\arg(1 - \tilde{k}(s))]_{s=-\infty}^{s=\infty} = 0. \tag{2.3}$$

The integer v occurring in (2.3) is called the *index* of Eq. (WH). We shall now put the equation

$$x(t) = h(t) + \int_0^\infty k(t-s)x(s) \, ds, \qquad t \in R_+, \tag{WH}$$

in the form

$$x(t) = h(t) + \int_R k(t-s)x(s) \, ds, \qquad t \in R, \tag{2.4}$$

in order to make use of the results we established in Section 2.14 for convolution equations on the real line.

Assume there exists a (measurable) solution of Eq. (WH). If we define

$$b(t) = \begin{cases} -\int_0^\infty k(t-s)\,x(s)\,ds, & t < 0, \\ 0, & t \geq 0, \end{cases} \tag{2.5}$$

$$x(t) = h(t) = 0, \qquad\qquad\qquad t < 0, \tag{2.6}$$

then Eq. (WH) can be written in the form (2.4), with $h(t) + b(t)$ instead of $h(t)$ in the right-hand side:

$$x(t) = h(t) + b(t) + \int_R k(t-s)x(s)\,ds, \qquad t \in R. \tag{2.7}$$

Indeed, (2.7) obviously reduces to (WH) for $t > 0$ and, according to (2.5), (2.6) it is also verified for $t < 0$.

Let us first consider Eq. (WH) in the space $E = L(R_+, \mathscr{C})$. If there exists a solution $x \in E$, then (2.7) has a solution in $L(R, \mathscr{C})$. Since h and b are also in $L(R, \mathscr{C})$, we can apply the Fourier transform to both sides of (2.7) and obtain

$$[1 - \tilde{k}(s)]\tilde{x}(s) = \tilde{h}(s) + \tilde{b}(s), \qquad s \in R. \tag{2.8}$$

Conditions (2.2) and (2.3) guarantee that the function $G(s) = [1 - \tilde{k}(s)]^{-1}$ has a unique canonical factorization

$$G(s) = G_+(s)G_-(s), \qquad s \in R, \tag{2.9}$$

with

$$G_+(s) = 1 + \int_0^\infty \gamma_1(t)e^{ist}\,dt, \qquad \operatorname{Im} s \geq 0, \tag{2.10}$$

and

$$G_-(s) = 1 + \int_0^\infty \gamma_2(t)e^{-ist}\,dt, \qquad \operatorname{Im} s \leq 0, \tag{2.11}$$

where $\gamma_1(t), \gamma_2(t) \in L(R_+, \mathscr{C})$. These facts follow from Theorem 1.4.1. Moreover, $G_+(s) \neq 0$ for $\operatorname{Im} s \geq 0$ and $G_-(s) \neq 0$ for $\operatorname{Im} s \leq 0$.

At this point, some auxiliary comments are necessary. Let us consider the class V of functions that are representable in the form

$$\varphi(s) = c + \int_R h(t)e^{ist}\,dt, \qquad s \in R, \tag{2.12}$$

with $c \in \mathscr{C}$ and $h(t) \in L(R, \mathscr{C})$. it is easy to see that V is a ring with respect to the usual operations of addition and multiplication. The norm is given by $\|\varphi\| = |c| + |h|_L$ (see Section 1.3).

We now define on V the projector P_+ by means of the formula

$$P_+\left(c + \int_R h(t)e^{ist}\,dt\right) = c + \int_0^\infty h(t)e^{ist}\,dt. \qquad (2.13)$$

The set $V_+ = P_+ V$ is obviously a subring of V. The notation V_- has a similar meaning. An element of V, V_+, or V_- is the Fourier transform of a function from $L(R, \mathscr{C})$, $L(R_+, \mathscr{C})$, or $L(R_-, \mathscr{C})$, respectively, if and only if it vanishes at infinity. Actually, the elements of V_+ are analytic functions in the half-plane $\operatorname{Im} s \geq 0$ and they can be regarded as Laplace transforms of functions from $L(R_+, \mathscr{C})$. A similar remark holds for V_-.

Now, let us remark that Eq. (2.8) can be written—according to (2.9)—in the form

$$\tilde{x}(s)/G_+(s) = G_-(s)\tilde{h}(s) + G_-(s)\tilde{b}(s), \qquad (2.14)$$

because both factors in the canonical factorization cannot vanish. From Theorem 1.3.6 there results easily that $1/G_+(s) \in V_+$. Taking into account that $\tilde{x} \in V_+$, $G_-(s) \in V_-$, and $\tilde{b}(s) \in V_-$, we obtain from (2.14) $\tilde{x}(s)/G_+(s) = P_+(G_-(s)\tilde{h}(s))$, which gives

$$\tilde{x}(s) = G_+(s)P_+(G_-(s)\tilde{h}(s)). \qquad (2.15)$$

Consequently, if Eq. (WH) has a solution in $L(R_+, \mathscr{C})$, its Fourier transform is given by (2.15). It follows then that such a solution is unique.

We are now going to prove the existence of the solution, starting from (2.15). In other words, if conditions (2.1), (2.2), and (2.3) are satisfied, Eq. (WH) has at least one solution in $E = L(R_+, \mathscr{C})$. Indeed, under our assumptions, (2.15) uniquely defines a function that is the Fourier transform of a function $x(t) \in L(R_+, \mathscr{C})$. The right-hand side of (2.15) vanishes at infinity because $\tilde{h}(s)$ does. We will show that this function $x(t)$ satisfies Eq. (WH).

From (2.15) we obtain $\tilde{x}(s)/G_+(s) = P_+(G_-(s)\tilde{h}(s))$, which means that $\tilde{x}(s)/G_+(s) = G_-(s)\tilde{h}(s) + \lambda(s)$, with $\lambda(s) \in V_-$ and $\lambda(\infty) = 0$. We get further $[1 - \tilde{k}(s)]\tilde{x}(s) = \tilde{h}(s) + \mu(s)$, where $\mu(s) \doteq \lambda(s)/G_-(s)$. Since $1/G_-(s) \in V_-$, there results $\mu(s) \in V_-$ and $\mu(\infty) = 0$. Therefore, $P_+([1 - \tilde{k}(s)]\tilde{x}(s)) = \tilde{h}(s)$, which leads to

$$\int_0^\infty e^{ist}\left[x(t) - \int_0^\infty k(t-u)x(u)\,du\right]dt = \int_0^\infty e^{ist}h(t)\,dt. \qquad (2.16)$$

From (2.16) we obtain

$$\int_0^\infty e^{ist}\left[x(t) - \int_0^\infty k(t-u)x(u)\,du - h(t)\right]dt = 0 \qquad \text{a.e.,}$$

a condition equivalent to Eq. (WH) if we consider that the only function whose Fourier transform vanishes identically is the null function (of course, almost everywhere).

To summarize the above discussion, we shall state the following partial result: If Eq. (WH) satisfies the conditions (2.1)–(2.3), then there exists a unique solution $x(t) \in L(R_+, \mathscr{C})$ for any $h(t) \in L(R_+, \mathscr{C})$.

We shall now find the representation of the solution by means of the resolvent kernel associated with $k(t)$. Let us notice that there results

$$P_+(G_-(s)\tilde{h}(s)) = \int_0^\infty e^{ist}g(t)\,dt, \qquad (2.17)$$

with

$$g(t) = h(t) + \int_0^\infty \gamma_2(-t+u)h(u)\,du, \qquad t \in R_+, \qquad (2.18)$$

considering (2.6). From (2.15), (2.17), and (2.10), one obtains

$$x(t) = g(t) + \int_0^\infty \gamma_1(t-u)g(u)\,du, \qquad t \in R_+, \qquad (2.19)$$

if we agree to define $g(t) = 0$ for $t < 0$. Finally, since both formulas (2.18) and (2.19) involve the values of $\gamma_1(t)$ and $\gamma_2(t)$ for negative t, we agree to consider as usual $\gamma_1(t) = \gamma_2(t) = 0$ for $t < 0$. Equations (2.18) and (2.19) lead to the formula

$$x(t) = h(t) + \int_0^\infty \gamma(t,s)h(s)\,ds, \qquad t \in R_+, \qquad (2.20)$$

where $\gamma(t, s)$ is given by

$$\gamma(t,s) = \gamma_1(t-s) + \gamma_2(s-t) + \int_0^\infty \gamma_1(t-u)\gamma_2(s-u)\,du \qquad (2.21)$$

for $0 \leq t, s < \infty$. Actually, the upper limit of the integral occurring in (2.21) can be replaced by $\min(t, s)$.

From the Remark 2 to Theorem 1.1 we derive the fact that $\gamma(t, s)$ satisfies conditions (a), (b), and (c) from the preceding section.

Before passing to the discussion of Eq. (WH) in the case where the space E is a function space other than $L(R_+, \mathscr{C})$, we shall find the integral equations for the resolvent kernel $\gamma(t, s)$. These equations will allow us to prove that Eq. (WH) has a unique solution $x(t) \in E$ for any $h \in E$, if conditions (2.1)–(2.3) are verified.

We claim that $\gamma(t, s)$ satisfies

$$\gamma(t,s) = k(t-s) + \int_0^\infty k(t-u)\gamma(u,s)\,du, \qquad (2.22)$$

and

$$\gamma(t, s) = k(t - s) + \int_0^\infty k(u - s)\gamma(t, u)\, du. \tag{2.23}$$

We notice that these equations should be understood as follows: if one of the variables is fixed, then they are verified almost everywhere with respect to the other. To obtain (2.22), for instance, we have to substitute $x(t)$ given by (2.20) into Eq. (WH). One obtains

$$\int_0^\infty \left[\gamma(t, s) - k(t - s) - \int_0^\infty k(t - u)\gamma(u, s)\, du\right] h(s)\, ds = 0,$$

which implies (2.22), due to the fact that $h(t) \in L(R_+, \mathscr{C})$ is arbitrary. Equation (2.23) can be obtained in the same manner, if we substitute $h(t)$ from (WH) in the formula (2.20).

In order to solve Eq. (WH) for an arbitrary space E occurring in the statement of Theorem 1.1 [i.e., M, L^p ($1 \le p \le \infty$), C, C_0], we shall use the formula (2.20). More precisely, we shall show that (2.20) gives a solution of the Eq. (WH) belonging to E, for any $h(t) \in E$. Of course, it suffices to consider the case $E = M(R_+, \mathscr{C})$, this one being the richest among the spaces listed above. Afterwards, the existence of a solution in any space E can be derived from Theorem 1.1.

That (2.20) defines a function $x(t) \in M(R_+, \mathscr{C})$ for any $h(t) \in M(R_+, \mathscr{C})$, one can see from Theorem 1.1. By direct calculation one sees that this function verifies Eq. (WH) a.e. on R_+. The key to the proof is the fact that $\gamma(t, s)$ satisfies the integral equations (2.22) and (2.23). The uniqueness follows also from the existence of $\gamma(t, s)$, and the proof is very similar to that given in Theorem 2.14.1.

Summing up the conclusions of the above discussion concerning (WH), we can state the following basic result.

Theorem 2.1   Consider Eq. (WH) and assume that $k(t)$ satisfies the conditions (2.1)–(2.3). Let E be any of the spaces $M(R_+, \mathscr{C})$, $L^p(R_+, \mathscr{C})$ ($1 \le p \le \infty$), $C(R_+, \mathscr{C})$, $C_\ell(R_+, \mathscr{C})$, $C_0(R_+, \mathscr{C})$. Then for any $h \in E$, Eq. (WH) has a unique solution $x \in E$. This solution is given by the formula (2.20), with $\gamma(t, s)$ constructed as above.

Remark   From (2.21) there results $\gamma(t, 0) = \gamma_1(t)$, $\gamma(0, t) = \gamma_2(s)$, $0 \le t$, $s < +\infty$. Since $\gamma_1(t)$ and $\gamma_2(s)$ completely determine $\gamma(t, s)$, we see that $\gamma(t, s)$ is known as soon as we know its values on the boundary of the domain of definition.

4.3   Equation (WH) with Positive Index

The aim of this section is the discussion of Eq. (WH) and of the corresponding homogeneous equation

$$y(t) = \int_0^\infty k(t - s)y(s)\,ds, \qquad t \in R_+,\tag{3.1}$$

under assumptions (2.1), (2.2), and

$$v = -\mathrm{ind}[1 - \tilde{k}(s)] > 0.\tag{3.2}$$

Under these assumptions, there exist infinitely many regular factorizations

$$[1 - \tilde{k}(s)]^{-1} = G_+(s)G_-(s), \qquad s \in R,\tag{3.3}$$

with $G_-(s) \neq 0$ for $\mathrm{Im}\, s \leq 0$. Concerning $G_+(s)$, it has exactly v zeros in the half-plane $\mathrm{Im}\, s > 0$ (see Theorem 1.4.2 and the remark to this theorem). Of course, the possibility of arbitrarily choosing these v zeros in the half-plane $\mathrm{Im}\, s > 0$ explains the existence of infinitely many regular factorizations for $[1 - \tilde{k}(s)]^{-1}$.

In the preceding section it was shown that under assumptions (2.1), (2.2), and $v = 0$, the formula

$$\tilde{x}(s) = G_+(s)P_+(G_-(s)\tilde{h}(s))\tag{3.4}$$

defines the Fourier transform of the solution belonging to $L(R_+, \mathscr{C})$, for any $h \in L(R_+, \mathscr{C})$.

We shall now prove that (3.4) also defines the Fourier transform of a solution of Eq. (WH), no matter how we choose the regular factorization (3.3). First, we assume that $E = L(R_+, \mathscr{C})$.

Indeed, for any $h \in L(R_+, \mathscr{C})$, the formula (3.4) gives a function $\tilde{x} \in V_+$, because $G_+(s)$ and $G_-(s)$ can be represented by the formulas (2.10) and (2.11) (see Theorem 1.4.2). Of course, $\tilde{x}$ determines a unique $x \in L(R_+, \mathscr{C})$. From (3.4) we obtain

$$\tilde{x}(s)/G_+(s) = G_-(s)\tilde{h}(s) + \lambda(s),$$

with $\lambda(s) \in V_-$ and $\lambda(\infty) = 0$. Since $1/G_-(s) \in V_-$, we have further

$$[1 - \tilde{k}(s)]\tilde{x}(s) = \tilde{h}(s) + \mu(s),$$

where $\mu(s) = \lambda(s)/G_-(s) \in V_-$ and $\mu(\infty) = 0$.

Therefore, we can write

$$P_+[(1 - \tilde{k}(s))\tilde{x}(s)] = \tilde{h}(s),\tag{3.5}$$

which leads to the conclusion that $x(t)$ satisfies Eq. (WH). The argument was developed in the preceding section.

Since $G_+(s)$ and $G_-(s)$ are representable by (2.10) and (2.11), with convenient $\gamma_1(t)$, $\gamma_2(t) \in L(R_+, \mathscr{C})$, we again find for $x(t)$ the formula

$$x(t) = h(t) + \int_0^\infty \gamma(t, s)h(s)\,ds, \qquad t \in R_+, \tag{3.6}$$

with $\gamma(t, s)$ given by

$$\gamma(t, s) = \gamma_1(t - s) + \gamma_2(s - t) + \int_0^\infty k(t - u)\gamma(s - u)\,du, \tag{3.7}$$

where $\gamma_1(t) = \gamma_2(t) = 0$ for $t < 0$. Of course, $\gamma(t, s)$ satisfies the integral equation

$$\gamma(t, s) = k(t - s) + \int_0^\infty k(t - u)\gamma(u, s)\,du,$$

which allows us to use formula (3.6) in constructing solutions for Eq. (WH) in any space E.

We can state the following result.

Theorem 3.1 If $k(t)$ satisfies conditions (2.1) and (2.2), and the index is positive, then for any space E and $h \in E$, there exist infinitely many solutions belonging to E of the Eq. (WH). They are given by formula (3.6), with $\gamma(t, s)$ constructed starting from an arbitrary regular factorization of the form (3.3).

The existence of infinitely many solutions for the Eq. (WH), corresponding to a given $h(t)$, implies the existence of nonzero solutions for the Eq. (3.1). We are now going to clarify how many solutions exist for the homogeneous Eq. (3.1). In order to formulate the answer, we need the following definition.

Let L be a finite dimensional function space whose elements are complex-valued functions on R_+. A basis $\{\varphi_0, \varphi_1, \ldots, \varphi_{\nu-1}\}$ is called a D-basis if the following conditions hold:

1. $\varphi_k(t)$, $k = 0, 1, \ldots, \nu - 1$, are absolutely continuous on any compact interval of R_+;
2. $\varphi_{k+1}(t) = \dot{\varphi}_k(t)$, $\varphi_k(0) = 0$, $k = 0, 1, \ldots, \nu - 2$;
3. $\varphi_{\nu-1}(0) \neq 0$.

The following result clarifies the properties of Eq. (3.1) when $\nu > 0$ and $E = L(R_+, \mathscr{C})$.

Theorem 3.2 Consider Eq. (3.1) under assumptions (2.1), (2.2), and $\nu > 0$. Then the dimension of the space L of all solutions of Eq. (3.1) belonging to $L(R_+, \mathscr{C})$ is ν, and there exists a D-basis whose functions belong to $C_0(R_+, \mathscr{C})$ and are absolutely continuous.

Proof   Consider the regular factorization (3.3) corresponding to the case when $G_+(s)$ has $s = i$ as a zero of multiplicity v.

We shall prove first that

$$G_k(s) = i^k G_+(s)/(s - i)^k, \qquad k = 1, 2, \ldots, v, \tag{3.8}$$

can be represented in the form

$$G_k(s) = \int_0^\infty e^{ist} g_k(t)\, dt, \qquad k = 1, 2, \ldots, v, \tag{3.9}$$

with $g_k(t)$ absolutely continuous and belonging to $C_0(R_+, \mathscr{C})$. Moreover, the $g_k(t)$ satisfy the following conditions:

$$\dot{g}_1(t) - g_1(t) = \gamma_1(t), \qquad g_1(0) = 1,$$
$$\dot{g}_k(t) - g_k(t) = g_{k-1}(t), \qquad g_k(0) = 0, \tag{3.10}$$
$$k = 2, 3, \ldots, v,$$

where $\gamma_1(t) \in L(R_+, \mathscr{C})$ is the function occurring in the formula

$$G_+(s) = 1 + \int_0^\infty \gamma_1(t) e^{ist}\, dt. \tag{3.11}$$

We claim that

$$g_1(t) = -e^t \int_t^\infty \gamma_1(s) e^{-s}\, ds = -\int_t^\infty \gamma_1(s) e^{t-s}\, ds, \qquad t \in R_+, \tag{3.12}$$

satisfies the required conditions. Since $g_1(t)$ is a convolution product, (3.9) follows easily for $k = 1$ if we take into account that

$$i/(s - i) = -\int_{-\infty}^0 e^{ist} e^t\, dt.$$

The equation $\dot{g}_1(t) - g_1(t) = \gamma_1(t)$ can be derived by a direct calculation starting from (3.12). Finally, we get from (3.12),

$$g_1(0) = -\int_0^\infty \gamma_1(s) e^{-s}\, ds. \tag{3.13}$$

But $G_1(i) = 0$, and this means that

$$1 + \int_0^\infty \gamma_1(t) e^{-t}\, dt = 0. \tag{3.14}$$

From (3.13) and (3.14) we get $g_1(0) = 1$. Since

$$|g_1(t)| \le \int_t^\infty |\gamma_1(s)|\, ds, \qquad t \in R_+, \tag{3.15}$$

we see that $g_1(t) \in C_0(R_+, \mathscr{C})$. Hence, everything is proved for $k = 1$. The proof follows the same lines for $k = 2, 3, \ldots, v$. It is useful to remark that

$$G_k(s) = iG_{k-1}(s)/(s - i), \qquad h = 2, 3, \ldots, v.$$

This leads to

$$g_k(t) = -\int_t^\infty g_{k-1}(s)e^{t-s}\, ds, \qquad t \in R_+, \tag{3.16}$$

for $h = 2, 3, \ldots, v$. From (3.16) we obtain

$$g_k(0) = -\int_0^\infty g_{k-1}(s)e^{-s}\, ds. \tag{3.17}$$

On the other hand, (3.9) yields

$$0 = G_k(i) = \int_0^\infty g_k(t)e^{-t}\, dt, \tag{3.18}$$

$k = 1, 2, \ldots, v - 1$. If we compare (3.17) and (3.18), there results $g_k(0) = 0$ for $k = 2, 3, \ldots, v$. The fact that the $g_k(t)$, $k = 1, 2, \ldots, v$, are absolutely continuous on R_+ is the consequence of the formulas (3.12) and (3.16). To show that the $g_k(t)$ belong to $L(R, \mathscr{C})$, one can proceed as follows. For instance, $g_1(t)$ is the convolution product of two functions from $L(R, \mathscr{C})$. The first one is $\gamma_1(t)$, with $\gamma_1(t) = 0$ for $t < 0$, and the second function equals $-e^t$ for $t \leq 0$ and is identically zero for $t > 0$.

Let us remark now that formula (3.3) can be written as

$$[1 - \tilde{k}(s)]G_+(s) = 1/G_-(s),$$

which shows that

$$[1 - \tilde{k}(s)]G_k(s) = i^k/G_-(s)(s - i)^k, \tag{3.19}$$

if we consider (3.8). Inasmuch as $1/G_-(s) \in V_-$ and $(s - i)^{-k} \in V_-$ vanish at infinity, we see that the right-hand side in (3.19) belongs to V_- and vanishes at infinity. Therefore

$$P_+[(1 - \tilde{k}(s))G_k(s)] \equiv 0, \qquad k = 1, 2, \ldots, v.$$

The preceding equations mean that

$$g_k(t) - \int_0^\infty k(t - s)g_k(s)\, ds = 0, \qquad t \in R_+, \quad k = 1, 2, \ldots, v.$$

Therefore, the functions $g_k(t)$, $k = 1, 2, \ldots, v$, are solutions of Eq. (3.1).

In order to obtain a D-basis for the space L, we will consider the following functions:

$$\varphi_0(t) = g_\nu(t),$$

$$\varphi_1(t) = \varphi_0'(t) = g_\nu(t) + g_{\nu-1}(t) \tag{3.20}$$

$$\vdots$$

$$\varphi_{\nu-1}(t) = \varphi_0^{(\nu-1)}(t) = \sum_{k=0}^{\nu-1} \binom{\nu-1}{k} g_{\nu-k}(t).$$

Condition (3) from the definition of a D-basis follows from

$$\varphi_{\nu-1}(0) = g_1(0) = 1,$$

while conditions (1) and (2) are obviously verified. It remains to show that the system $\{\varphi_0, \varphi_1, \ldots, \varphi_{\nu-1}\}$ consists of linearly independent functions and that any solution of (3.1) belonging to $L(R_+, \mathscr{C})$ can be expressed as a linear combination of the solutions $\varphi_0, \varphi_1, \ldots, \varphi_{\nu-1}$.

Since $\varphi_k \in C_0(R_+, \mathscr{C})$, $k = 0, 1, \ldots, \nu - 1$, we can write

$$\varphi_0^{(k-1)}(t) = \int_0^t \varphi_0^{(k)}(s)\, ds$$

$$= -\int_t^\infty \varphi_0^{(k)}(s)\, ds, \qquad k = 1, 2, \ldots, \nu - 1, \tag{3.21}$$

and

$$\varphi_0^{(\nu-1)}(t) = 1 + \int_0^t \varphi_0^{(\nu)}(s)\, ds = -\int_t^\infty \varphi_0^{(\nu)}(s)\, ds. \tag{3.22}$$

Taking into account (3.20), (3.21), and the fact that $\varphi_k(0) = 0$, $k = 0, 1, \ldots, \nu - 2$, we obtain

$$\tilde{\varphi}_k(s) = (-is)^k \tilde{\varphi}_0(s), \qquad k = 1, 2, \ldots, \nu - 1. \tag{3.23}$$

But $\tilde{\varphi}_0(s) = G_\nu(s) = i^\nu G_+(s)/(s-i)^\nu$ does not vanish for $\mathrm{Im}\, s > 0$. Consequently, the system $\{\varphi_0, \varphi_1, \ldots, \varphi_{\nu-1}\}$ consists of linearly independent functions [otherwise, the functions $\tilde{\varphi}_k(s)$, $k = 0, 1, \ldots, \nu - 1$, would be linearly dependent, which is obviously impossible].

Now let $\varphi(t)$ be an arbitrary function belonging to $L(R_+, \mathscr{C})$. We can choose ν constants $c_0, c_1, \ldots, c_{\nu-1}$ with the property that

$$\omega(t) = \varphi(t) - \sum_{k=0}^{\nu-1} c_k \varphi_k(t)$$

is such that $\tilde{\omega}(s)$ has a zero of multiplicity ν for $s = i$. In other words, $\tilde{\omega}^{(k)}(i) = 0$, $k = 0, 1, 2, \ldots, \nu - 1$. Obviously, these constants are such that the polynomial $\sum_{k=0}^{\nu-1} c_k(-is)^k$ coincides with the sum of the first ν terms in the Taylor's series $\tilde{\varphi}(s)/\tilde{\varphi}_0(s) = \sum_{k=0}^\infty a_k(s-i)^k$. Indeed,

$$\tilde{\omega}(s) = \tilde{\varphi}(s) - \tilde{\varphi}_0(s) \sum_{k=0}^{\nu-1} c_k(-is)^k.$$

If $\varphi \in L(R_+, \mathscr{C})$ is a solution of (3.1), then ω is also a solution of the same equation. This yields $P_+([1 - \tilde{k}(s)]\tilde{\omega}(s)) = 0$, which implies $[1 - \tilde{k}(s)]\tilde{\omega}(s) = \lambda(s)$, with $\lambda(s) \in V_-$ and $\lambda(\infty) = 0$. Considering (3.3), we derive $\tilde{\omega}(s)/G_+(s) = \lambda(s)G_-(s)$. Since $G_+(s)$ has a single zero of multiplicity v at $s = i$, we can assert that $\tilde{\omega}(s)/G_+(s) \in V_+$. Indeed, from (3.8) we obtain

$$G_+(s) = i^{-k}G_v(s)/(s - i)^v.$$

Hence $\tilde{\omega}(s)/G_+(s) = [i^v \tilde{\omega}(s)/(s - i)^v]/G_v(s)$ with $G_v(s) \in V_+$ and $G_v(s) \neq 0$ for Im $s \geq 0$. Therefore, $\tilde{\omega}(s)/G_+(s) \in V_+$. On the other hand, $\lambda(s)$, $G_-(s) \in V_-$ and vanish at infinity. We obtain $\tilde{\omega}(s) = \lambda(s) \equiv 0$, which means that $\omega(t) \equiv 0$. In other words, $\varphi(t)$ can be expressed as a linear combination of the functions $\varphi_0(t), \ldots, \varphi_{v-1}(t)$. The proof of Theorem 3.2 is complete.

Remark Theorem 3.2 is concerned with the homogeneous equation (3.1) in the space $L(R_+, \mathscr{C})$. In the next section we shall prove that the solutions belonging to $L(R_+, \mathscr{C})$ are the only solutions of Eq. (3.1) in any space E, where E stands for any of the spaces M, L^p $(1 \leq p \leq \infty)$, C, C_ℓ, C_0. In order to obtain this result, we need further information on Eq. (WH).

4.4   Equation (WH) with Negative Index

Consider the transpose of Eq. (WH), namely,

$$z(s) = f(s) + \int_0^\infty k(t - s)z(t)\,dt, \qquad s \in R_+. \tag{4.1}$$

If (WH) has negative index $v = -\text{ind}[1 - \tilde{k}(s)] < 0$, then Eq. (4.1) has positive index $|v| = \text{ind}[1 - \tilde{k}(s)] = -v > 0$. Indeed, this follows easily from the fact that the Fourier transform of $k(-t)$ is $\tilde{k}(-s)$. Therefore, Theorem 3.1 allows us to state that Eq. (4.1) has infinitely many solutions in any space E, and for arbitrary $f \in E$. Moreover, there exist infinitely many resolvent kernels $\gamma(t, s)$, with the property that any function of the form

$$z(s) = f(s) + \int_0^\infty \gamma(t, s)f(t)\,dt, \qquad s \in R_+, \tag{4.2}$$

with $f \in E$, is a solution of Eq. (4.1) and belongs to E. In order to clarify the last statement, it suffices to start from the regular factorization

$$[1 - \tilde{k}(-s)]^{-1} = G_-(-s)G_+(-s),$$

and to take into account the remark to Theorem 1.4.2.

We shall now establish an auxiliary result that plays a significant role in conducting further our discussion.

Lemma 4.1   Assume we are given $h(t) \in M(R_+, \mathscr{C})$ and let $g(t) \in L(R_+, \mathscr{C})$ be such that $\dot{g}(t) \in L(R_+, \mathscr{C})$. Then $h(t)g(t) \in L(R_+, \mathscr{C})$ and

$$\lim_{t \to 0} \int_0^\infty h(s)g(t+s)\,ds = \int_0^\infty h(s)g(s)\,ds. \tag{4.3}$$

Proof   We set

$$H(t) = \int_0^\infty h(s)g(t+s)\,ds, \qquad t > 0. \tag{4.4}$$

It is known (Section 2.7) that $H(t)$ is defined for almost all $t \in R_+$ and belongs to M. Therefore, we can consider the derivative $\dot{H}(t)$ in the sense of distributions (see Yosida [1, Chapter I, §8]). It can be readily seen that a.e. on R_+

$$\dot{H}(t) = \int_0^\infty h(s)\dot{g}(t+s)\,ds, \tag{4.5}$$

which shows that $\dot{H}(t) \in M$. Moreover, we can assert that $H(t)$ is absolutely continuous on any compact interval of R_+.

A similar argument holds with respect to the function

$$H_1(t) = \int_0^\infty |h(s)|\,|g(t+s)|\,ds. \tag{4.6}$$

Since $|h(s)g(t+s)| \to |h(s)g(s)|$ as $t \to 0$ for almost all $s \in R_+$, from Fatou's lemma and (4.6) we get

$$\int_0^\infty |h(s)g(s)|\,ds \le H_1(0). \tag{4.7}$$

Hence, $h(t)g(t) \in L(R_+, \mathscr{C})$.

Let us now show that (4.3) holds true. It suffices to prove that

$$\lim_{t \to 0} \int_0^\infty h(s)[g(t+s) - g(s)]\,ds = 0. \tag{4.8}$$

We have $g(t+s) - g(s) = \int_0^t \dot{g}(s+u)\,du$ and

$$\int_0^t du \int_0^\infty |h(s)|\,|\dot{g}(s+u)|\,ds < +\infty$$

for any $t > 0$. Therefore,

$$\int_0^\infty h(s)[g(t+s) - g(s)]\,ds = \int_0^t du \int_0^\infty h(s)\dot{g}(s+u)\,ds$$

$$= \int_0^t \dot{H}(u)\,du \to 0 \qquad \text{as} \quad t \to 0.$$

Consequently, (4.8) and (4.3) hold, which proves Lemma 4.1.

The following result establishes a basic property of the homogeneous equation

$$y(t) = \int_0^\infty k(t - s)y(s)\, ds, \qquad t \in R_+ . \tag{3.1}$$

Theorem 4.1 Assume that $v = -\mathrm{ind}[1 - \tilde{k}(s)] \leq 0$. Then Eq. (3.1) has only the solution $y(t) \equiv 0$ in the space E, where E stands for any of the spaces M, L^p $(1 \leq p \leq \infty)$, C, C_ℓ, C_0.

Proof It is obvious that it suffices to discuss only the case $E = M$.

Let $y(t) \in M$ be a nontrivial solution of Eq. (3.1). We can construct an absolutely continuous function $f(t)$, with $f(t) \equiv 0$ for $t \geq T_0 > 0$, such that

$$\int_0^\infty f(t)y(t)\, dt = 1. \tag{4.9}$$

For instance, we can choose $f(t)$ to coincide with a polynomial on $[0, T_0]$. From (4.2) we see that $z(t) \in L(R_+, \mathscr{C})$ and, therefore,

$$\int_0^\infty y(s)f(u + s)\, ds = \int_0^\infty y(s)\left[z(u + s) - \int_0^\infty k(t - u - s)z(t)\, dt \right] ds$$

$$= \int_0^\infty y(s)z(u + s)\, ds - \int_0^\infty z(t)\, dt \int_0^\infty k(t - u - s)y(s)\, ds$$

$$= \int_0^\infty y(s)z(u + s)\, ds - \int_0^\infty z(t)y(t - u)\, dt = 0.$$

The legitimacy of these operations is a consequence of the fact that for almost all $u \in R_+$, the function $|k(t - u - s)y(s)z(t)|$ is integrable on $R_+ \times R_+$. Indeed, the iterated integral (first with respect to s and then with respect to t) is convergent. From $\int_0^\infty y(s)f(u + s)\, ds = 0$ and the Lemma 4.1, we obtain $\int_0^\infty y(s)f(s)\, ds = 0$, which contradicts (4.9). Theorem 4.1 is thereby proven.

Remark Theorem 4.1 completes, in a certain sense, Theorem 2.1. For $v = 0$, the property stated in Theorem 4.1 can be derived from Theorem 2.1. Indeed, if (WH) has a unique solution $x \in E$ for any $h \in E$, then (3.1) has only the solution $y(t) \equiv 0$ in any space E.

In order to complete the discussion of Eq. (WH), we have to consider the case

$$v = -\mathrm{ind}[1 - \tilde{k}(s)] < 0. \tag{4.10}$$

Theorem 4.2 Consider Eq. (WH) under assumptions (2.1), (2.2), and
(4.10). Let E be any of the spaces $M(R_+, \mathscr{C})$, $L^p(R_+, \mathscr{C})$ ($1 \leq p \leq \infty$),
$C(R_+, \mathscr{C})$, $C_\ell(R_+, \mathscr{C})$, $C_0(R_+, \mathscr{C})$. Then given $h \in E$, Eq. (WH) has solutions
$x \in E$ if and only if

$$\int_0^\infty h(t)\psi_j(t)\, dt = 0, \qquad j = 0, 1, \ldots, |v| - 1, \tag{4.11}$$

where $\{\psi_0, \psi_1, \ldots, \psi_{|v|-1}\}$ is a D-basis of the space of all solutions belonging
to $L(R_+, \mathscr{C})$ of the homogeneous transpose equation

$$\psi(s) = \int_0^\infty k(t - s)\psi(t)\, dt, \qquad s \in R_+. \tag{4.12}$$

If (4.11) holds, then (WH) has a unique solution $x \in E$.

Proof In order to prove the necessity part, we remark that the existence
of the integrals occurring in (4.11) follows from Theorem 3.2, because the
index of Eq. (4.12) is $|v| > 0$, and from the Lemma 4.1. Now, if (WH) has a
solution $x \in E$ for given $h \in E$, multiplying both sides of (WH) by $\psi_j(t)$, where
the $\psi_j(t)$, $j = 0, 1, \ldots, |v| - 1$, form a D-basis of L-solutions for (4.12), and
integrating we get

$$\int_0^\infty x(t)\psi_j(t)\, dt = \int_0^\infty h(t)\psi_j(t)\, dt + \int_0^\infty \psi_j(t)\, dt \int_0^\infty k(t - s)x(s)\, ds$$

$$= \int_0^\infty h(t)\psi_j(t)\, dt + \int_0^\infty x(s)\, ds \int_0^\infty k(t - s)\psi_j(t)\, dt$$

$$= \int_0^\infty h(t)\psi_j(t)\, dt + \int_0^\infty x(s)\psi_j(s)\, ds.$$

This obviously implies conditions (4.11).

Let us now prove that conditions (4.11) are also sufficient. As seen above,
if $v < 0$, Eq. (4.1) has infinitely many solutions in E for any $f \in E$. These
solutions are given by the formula (4.2), with $\gamma(t, s)$ constructed as indicated
in the preceding section. We will show that the function

$$x(t) = h(t) + \int_0^\infty \gamma(t, s)h(s)\, ds, \qquad t \in R_+, \tag{4.13}$$

with $\gamma(t, s)$ from (4.2) and $h(t)$ satisfying (4.11), is a solution of Eq. (WH).
For $h \in E$, we have $x \in E$.

Now, let $g(t) \in L(R_+, \mathscr{C})$ be arbitrarily chosen. Denote

$$f(s) = g(s) - \int_0^\infty k(t - s)g(t)\, dt, \qquad s \in R_+. \tag{4.14}$$

Then $f \in L(R_+, \mathscr{C})$ and

$$g_0(s) = f(s) + \int_0^\infty \gamma(t, s)f(t)\, dt, \qquad s \in R_+, \tag{4.15}$$

verifies (4.1). Of course, we cannot assert that $g_0(t) \equiv g(t)$ because the homogeneous equation has nontrivial solutions. All we can say is that

$$g_0(s) = g(s) + \sum_{k=0}^{|v|-1} c_k \psi_k(s), \tag{4.16}$$

where $c_k = c_k(g)$, $k = 0, 1, \ldots, |v| - 1$, are some functionals on $L(R_+, \mathscr{C})$. If we replace $f(s)$, given by (4.14), in (4.15) and compare the result with (4.16), we get

$$\int_0^\infty \left[\gamma(t, s) - k(t - s) - \int_0^\infty k(t - u)\gamma(u, s)\, du \right] g(t)\, dt = \sum_{k=0}^{|v|-1} c_k(g)\psi_k(s). \tag{4.17}$$

We shall prove that the functionals $c_k(g)$ can be expressed in the form

$$c_j(g) = \int_0^\infty \rho_j(t)g(t)\, dt, \qquad j = 0, 1, \ldots, |v| - 1, \tag{4.18}$$

with $\rho_j(t) \in E_0$, $j = 0, 1, \ldots, |v| - 1$, where E_0 denotes the intersection of all spaces E from the statement of the theorem. Indeed, let us consider some functions $\omega_j \in E_0$, $j = 0, 1, \ldots, |v| - 1$, such that

$$\int_0^\infty \omega_j(s)\psi_k(s)\, ds = \begin{cases} 1, & j = k, \\ 0, & j \neq k. \end{cases} \tag{4.19}$$

Let us now multiply both sides of (4.17) by $\omega_j(s)$ and integrate from 0 to ∞. Then we obtain (4.18), with

$$\rho_j(t) = \int_0^\infty \gamma(t, s)\omega_j(s)\, ds - \int_0^\infty k(t - u)\left[\omega_j(u) + \int_0^\infty \gamma(u, s)\omega_j(s)\, ds \right] du,$$

$j = 0, 1, \ldots, |v| - 1$. Inasmuch as $\omega_j \in E_0$, there results $\rho_j(t) \in E_0$. If we now substitute $c_j(g)$ from (4.18) in (4.17) and take into account that $g \in L(R_+, \mathscr{C})$ is arbitrary, we obtain for $0 \leq s, t < +\infty$

$$\gamma(t, s) = k(t - s) + \int_0^\infty k(t - u)\gamma(u, s)\, du + \sum_{j=0}^{|v|-1} \rho_j(t)\psi_j(s). \tag{4.20}$$

It is very easy now to prove that $x(t)$ given by (4.13), with $h \in E$, satisfies Eq. (WH). We multiply both sides in (4.20) by $h(s)$ and integrate from 0 to ∞ with respect to s. Taking into account conditions (4.11), there results

$$\int_0^\infty \gamma(t, s)h(s)\, ds = \int_0^\infty k(t - s)h(s)\, ds + \int_0^\infty h(s)\, ds \int_0^\infty k(t - u)\gamma(u, s)\, du.$$

But (4.13) gives

$$\int_0^\infty \gamma(t, s)h(s)\, ds = x(t) - h(t)$$

and the double integral can be transformed as follows:

$$\int_0^\infty h(s)\, ds \int_0^\infty k(t - u)\gamma(u, s)\, du = \int_0^\infty k(t - u)\, du \int_0^\infty \gamma(u, s)h(s)\, ds$$

$$= \int_0^\infty k(t - u)[x(u) - h(u)]\, du.$$

From the above relations one obtains

$$x(t) = h(t) + \int_0^\infty k(t - u)x(u)\, du,$$

i.e., $x(t)$ satisfies Eq. (WH).

The uniqueness follows from Theorem 4.1, and Theorem 4.2 is thereby proved.

We now go back to the homogeneous equation (3.1) in order to examine the existence of solutions in the space E, where E stands for any of the spaces M, L^p $(1 \le p \le \infty)$, C, C_ℓ, C_0. Of course, it remains to consider only the case $v > 0$ because Theorem 4.1 completely clarifies the case $v \le 0$. The result we want to prove can be stated as follows:

Theorem 4.3 If $k(t)$ satisfies (2.1), (2.2), and $v > 0$, then the solutions of Eq. (3.1) in the space E are the same, no matter how we choose E among the spaces M, L^p $(1 \le p \le \infty)$, C, C_ℓ, C_0.

Proof It suffices to show that any solution $y(t) \in M$ of Eq. (3.1) belongs to L. According to Theorem 4.2, Eq. (4.1) has a unique solution $z \in L$ for any $f \in L$ such that

$$\int_0^\infty f(t)\varphi_j(t)\, dt = 0, \qquad j = 0, 1, \ldots, v - 1, \tag{4.21}$$

where $\{\varphi_0, \varphi_1, \ldots, \varphi_{v-1}\}$ is a D-basis in the space of all L-solutions of Eq. (3.1). If we assume now that $y(t) \in M$ is a solution of Eq. (3.1) that is linearly independent of the solutions $\varphi_j(t)$, $j = 0, 1, \ldots, v - 1$, then there exists a function $f \in L$, with $\dot{f} \in L$, such that besides conditions (4.21) it also satisfies

$$\int_0^\infty f(t)y(t)\, dt \neq 0. \tag{4.22}$$

For almost all $u \in R_+$ we can write

$$\int_0^\infty f(t+u)y(t)\,dt = \int_0^\infty y(t)\left[z(t+u) - \int_0^\infty k(s-t-u)z(s)\,ds\right] dt$$

$$= \int_0^\infty y(t)z(t+u)\,dt - \int_0^\infty z(s)\,ds \int_0^\infty k(s-u-t)y(t)\,dt = 0,$$

if we take into account that the integral with respect to t in the double integral above equals $y(t+u)$. We can apply now Lemma 4.1 and obtain for $u \to 0$

$$\int_0^\infty f(t)y(t)\,dt = 0,$$

which contradicts (4.21). Theorem 4.3 is thus proved.

Theorem 4.3 completes the discussion of Eq. (WH) and of the equations related to it. Further properties related to Eq. (WH) can be found in the exercises at the end of this chapter.

4.5 Some Examples

As seen in the preceding sections, the problem of finding the resolvent kernel $\gamma(t, s)$ plays a significant role in solving and discussing Eq. (WH). We shall consider in this section some examples, in order to get a better idea of the difficulties one might encounter in the applications of the general results concerning Eq. (WH).

A. Consider Eq. (WH) and assume that

$$|k|_L = \int_R |k(t)|\,dt < 1. \tag{5.1}$$

Under condition (5.1), Eq. (WH) has a unique solution $x \in E$ for any $h \in E$. Moreover, this solution can be found by successive approximations.

Before proceeding to the construction of the kernel $\gamma(t, s)$, we remark that condition (5.1) implies $|\tilde{k}(s)| < 1$, $s \in R$. Consequently, (5.1) is stronger than (2.1). Inasmuch as $|1 - \tilde{k}(s)| \geq 1 - |\tilde{k}(s)| > 0$ for $s \in R$, there results also $\text{ind}[1 - \tilde{k}(s)] = 0$. Therefore, assumption (5.1) assures the existence and uniqueness of the solution of the Eq. (WH) in any space E.

An alternate proof of the existence and uniqueness can be obtained by means of the contraction mapping principle. Indeed, the operator T, defined by

$$(Tx)(t) = h(t) + \int_0^\infty k(t-s)x(s)\,ds, \qquad t \in R_+,$$

is a contraction ($\|T\|_E < 1$) in the space E, where E stands for any of the spaces M, L^p ($1 \leq p \leq \infty$), C, C_ℓ, C_0.

In order to construct $\gamma(t, s)$, we start from the integral equation (2.22),

$$\gamma(t, s) = k(t - s) + \int_0^\infty k(t - u)\gamma(u, s)\, du.$$

For $s = 0$, we obtain

$$\gamma(t, 0) = k(t) + \int_0^\infty k(t - u)\gamma(u, 0)\, du,$$

which easily leads to

$$\gamma(t, 0) = \sum_{n=0}^\infty k_n(t), \tag{5.2}$$

where

$$k_0(t) = k(t), \qquad k_{n+1}(t) = \int_0^\infty k(t - u)k_n(u)\, du, \qquad n \geq 0. \tag{5.3}$$

The convergence of the series (5.2) should be meant in the space $L(R_+, \mathscr{C})$. If we assume $k \in L \cap L^\infty$, then $k_n(t)$, $n \geq 1$, are continuous functions, and the convergence of the series occurring in (5.2) is uniform. Indeed, we can write

$$\left| \gamma(t, 0) - \sum_{n=0}^N k_n(t) \right| = \left| \int_0^\infty k(t - s) \sum_{n=N}^\infty k_n(s)\, ds \right|$$

$$\leq |k|_L \sum_{n=N}^\infty |k_n|_L.$$

Inasmuch as the series $\sum_{n=N}^\infty |k_n|_L$ converges, we get the uniform convergence of (5.2) in $C(R_+, \mathscr{C})$. More precisely, $\gamma(t, 0) - k(t) \in C(R_+, \mathscr{C})$, and

$$\gamma(t, 0) - k(t) = \sum_{n=1}^\infty k_n(t), \qquad t \in R_+,$$

the convergence being that of $C(R_+, \mathscr{C})$.

If we take $t = 0$ in the integral equation of the resolvent kernel, we can construct $\gamma(0, s)$ in the same manner as above.

The formula

$$\gamma(t, s) = \gamma(t - s, 0) + \gamma(0, s - t) + \int_0^\infty \gamma(t - u, 0)\gamma(0, s - u)\, du$$

yields the resolvent kernel associated with k.

B. Consider now Eq. (WH) under assumption that $\tilde{k}(s)$ is a rational function. Inasmuch as $\tilde{k}(s) \to 0$ as $s \to \infty$, there results

$$\tilde{k}(s) = \frac{b_0 s^{n-1} + b_1 s^{n-2} + \cdots + b_{n-1}}{a_0 s^n + a_1 s^{n-1} + \cdots + a_n}, \qquad a_0 \neq 0. \qquad (5.4)$$

From (5.4) we derive

$$1 - \tilde{k}(s) = \frac{(s - s_1)(s - s_2) \cdots (s - s_n)}{(s - \sigma_1)(s - \sigma_2) \cdots (s - \sigma_n)},$$

with $\operatorname{Im} \sigma_j \neq 0$, $j = 1, 2, \ldots, n$. In order to assure that $1 - \tilde{k}(s) \neq 0$, $s \in R$, it is necessary to assume that $\operatorname{Im} s_j \neq 0$, $j = 1, 2, \ldots, n$. Finally, taking into account a famous theorem in complex analysis (concerning the variation of the argument along a curve and the number of zeros and poles inside that curve), we see that there exists the same number of s_j and σ_j in the half-plane $\operatorname{Im} s > 0$ (and, of course, the same number in the half-plane $\operatorname{Im} s < 0$). Without loss of generality, we can assume $\operatorname{Im} s_j < 0$, $\operatorname{Im} \sigma_j < 0$ for $j = 1, 2, \ldots, m$, $m < n$, and $\operatorname{Im} s_j > 0$, $\operatorname{Im} \sigma_j > 0$, $j = m + 1, \ldots, n$. From $[1 - \tilde{k}(s)]^{-1} = G_+(s) G_-(s)$ and the conditions required in a canonical factorization, we obtain immediately

$$G_+(s) = \prod_{j=1}^{m} (s - \sigma_j)/(s - s_j), \qquad G_-(s) = \prod_{j=m+1}^{n} (s - \sigma_j)/(s - s_j).$$

If we assume now that the s_j, $j = 1, 2, \ldots, m$, are distinct, then

$$G_+(s) = 1 + \sum_{j=1}^{m} A_j/(s - s_j).$$

Taking into account the formula

$$G_+(s) = 1 + \int_0^{\infty} \gamma(t, 0) e^{ist}\, dt, \qquad \operatorname{Im} s \geq 0,$$

there results

$$\gamma(t, 0) = -i \sum_{j=1}^{m} A_j \exp(-is_j t).$$

If the s_j, $j = 1, 2, \ldots, m$, are not all distinct, then we get

$$\gamma(t, 0) = \sum_{j=1}^{r} P_j(t) \exp(-is_j t),$$

where the s_j, $j = 1, 2, \ldots, r < m$, represent all the distinct s_j, $j = 1, 2, \ldots, m$, and the $P_j(t)$ are certain polynomials.

C.   The last example we shall consider in this section regards Eq. (WH) with the kernel

$$k(t) = \mu/2\pi \, \cosh(t/2), \qquad t \in R, \tag{5.5}$$

μ being a parameter to be defined below.

The Fourier transform of the kernel is (see, for instance, Bateman and Erdelyi: "Tables of Integral Transforms," McGraw-Hill, 1954):

$$\tilde{k}(s) = \mu/\cosh \pi s, \qquad s \in R. \tag{5.6}$$

In order to assure the condition $1 - \tilde{k}(s) \neq 0$, $s \in R$, we have to assume $\mu \notin [1, \infty)$. Let $\mu = \cos \pi a$, with $0 < \operatorname{Re} a < 1$. Then

$$1 - \tilde{k}(s) = \frac{\cosh \pi s - \cos \pi a}{\cosh \pi s} = \frac{2 \sin[(\pi/2)(a + is)] \sin[(\pi/2)(a - is)]}{\cosh \pi s}. \tag{5 7}$$

If we consider the well-known equation

$$\Gamma(z)\Gamma(1 - z) = \pi/\sin \pi z,$$

where $\Gamma(z)$ denotes the Euler's function, then we can write

$$\sin \frac{\pi}{2}(a \pm is) = \frac{\pi}{\Gamma(\tfrac{1}{2}(a \pm is))\Gamma(1 - \tfrac{1}{2}(a \pm is))}$$

and

$$\cosh \pi s = \sin\left(\frac{\pi}{2} + i\pi s\right) = \frac{\pi}{\Gamma(\tfrac{1}{2} + is)\Gamma(\tfrac{1}{2} - is)}.$$

Now substituting $\sin([\pi/2](a \pm is)]$ and $\cosh \pi s$ from the above formulas into (5.7), we get

$$[1 - \tilde{k}(s)]^{-1} = (2\pi)^{-1} \frac{\Gamma(\tfrac{1}{2}(a + is))\Gamma(\tfrac{1}{2}(a - is))\Gamma(1 - \tfrac{1}{2}(a + is))\Gamma(1 - \tfrac{1}{2}(a - is))}{\Gamma(\tfrac{1}{2} + is)\Gamma(\tfrac{1}{2} - is)}.$$

Taking into account that $[1 - \tilde{k}(s)]^{-1} = G_+(s)G_-(s)$, $s \in R$, we have further

$$G_+(s) \frac{(2\pi)^{1/2}\Gamma(\tfrac{1}{2} - is)}{\Gamma(\tfrac{1}{2}(a - is))\Gamma(1 - \tfrac{1}{2}(a + is))}$$

$$= (2\pi)^{-1/2} \frac{\Gamma(\tfrac{1}{2}(a + is))\Gamma(1 - \tfrac{1}{2}(a - is))}{\Gamma(\tfrac{1}{2} + is)} \frac{1}{G_-(s)}$$

for any $s \in R$. But $1/\Gamma(z)$ is an entire function whose zeros are $0, -1, -2, \ldots$. If we consider that $0 < \operatorname{Re} a < 1$, then the reason we wrote the last equation becomes clear. Namely, it shows that the function

$$G_+(s) \frac{(2\pi)^{1/2}\Gamma(\tfrac{1}{2} - is)}{\Gamma(\tfrac{1}{2}(a - is))\Gamma(1 - \tfrac{1}{2}(a + is))},$$

which is holomorphic and does not vanish for Im $s \geq 0$, coincides for Im $s = 0$ with the function

$$(2\pi)^{-1/2} \frac{\Gamma(\frac{1}{2}(a + is))\Gamma(1 - \frac{1}{2}(a - is))}{\Gamma(\frac{1}{2} + is)} \frac{1}{G_-(s)},$$

which is holomorphic and does not vanish for Im $s \leq 0$. Therefore, we can write for Im $s \geq 0$

$$G_+(s) = G_-(-s) = (2\pi)^{-1/2} \frac{\Gamma(\frac{1}{2}(a - is))\Gamma(1 - \frac{1}{2}(a + is))}{\Gamma(\frac{1}{2} - is)} e^{\lambda(s)}, \quad (5.8)$$

where $\lambda(s) = -\lambda(-s)$ is an entire function. We shall now find $\lambda(s)$. The following asymptotic formula holds for $\Gamma(z)$, if $|\arg z| < \pi - \varepsilon, \varepsilon > 0$:

$$\ln \Gamma(z) = (z - \tfrac{1}{2}) \ln z - z + \tfrac{1}{2} \ln 2\pi + O(z^{-1})$$

as $z \to \infty$. This formula leads to

$$\ln \Gamma(\tfrac{1}{2} - is) = -is \ln(-is) + is + O(1),$$

$$\ln \Gamma(\tfrac{1}{2}(a - is)) = \tfrac{1}{2}(a - is - 1) \ln(-\tfrac{1}{2}is) + \tfrac{1}{2}is + O(1),$$

$$\ln \Gamma(1 - \tfrac{1}{2}(a + is)) = \tfrac{1}{2}(1 - a - is) \ln(-\tfrac{1}{2}is) + \tfrac{1}{2}is + O(1)$$

as $|s| \to \infty$ and Im $s > 0$. From (5.8) there results

$$\lambda(s) = \ln \Gamma(\tfrac{1}{2} - is) - \ln \Gamma(\tfrac{1}{2}(a - is)) - \ln \Gamma(1 - \tfrac{1}{2}(a + is)) + O(1),$$

if we take into account that $\ln G_+(s)$ tends to zero as $s \to \infty$. From the above formulas we obtain for Im $s > 0$

$$\lambda(s) = -is \ln 2 + O(1), \qquad s \to \infty. \quad (5.9)$$

But $\lambda(-s) = -\lambda(s)$ and this shows that (5.9) is valid for Im $s < 0$. According to Liouville's theorem, we have

$$\lambda(s) = -is \ln 2, \qquad s \in \mathscr{C}. \quad (5.10)$$

We can now write the final form for $G_+(s)$ and $G_-(s)$. There results

$$G_+(s) = G_-(-s) = (2\pi)^{-1/2} \frac{\Gamma(\frac{1}{2}(a - is))\Gamma(1 - \frac{1}{2}(a + is))}{\Gamma(\frac{1}{2} - is)} e^{-is \ln 2}. \quad (5.11)$$

Since $\gamma(t, 0)$ can be determined from

$$\gamma(t, 0) = (2\pi)^{-1} \int_R e^{-ist}[G_+(s) - 1] \, ds, \qquad t \in R_+, \quad (5.12)$$

it remains to calculate the integral occurring in the second member of (5.12). The meromorphic function $G_+(s) - 1$ has only simple poles in the half-plane

Im $s < 0$. These poles coincide with the zeros of the function $\cosh \pi s$ $- \cos \pi a = \cos \pi si - \cos \pi a$, i.e., they are $s_n = -i(a + 2n)$, $n = 0$, 1, 2, ..., and $s_n = i(a + 2n)$, $n = 1, -2, \ldots$. If we take into account that

$$G_+(s) - 1 = \frac{\cosh \pi s}{\cosh \pi s - \cos \pi a} \frac{1}{G_+(-s)} - 1,$$

then we can easily get the values of the residues corresponding to the poles s_n. We find

$$\frac{i \cot \pi a}{\pi G_+(-s_n)} e^{-is_n t} = i\left(\frac{2}{\pi}\right)^{1/2} \cot \pi a \frac{\Gamma(\frac{1}{2} + a + 2n)}{\Gamma(n + a)\Gamma(n + 1)} (2e^t)^{-(a + 2n)}$$

for $n = 0$, 1, 2, ..., and

$$-\frac{i \cot \pi a}{\pi G_+(-s_n)} e^{-is_n t} = -i\left(\frac{2}{\pi}\right)^{1/2} \cot \pi a \frac{\Gamma(\frac{1}{2} - a - 2n)}{\Gamma(-n)\Gamma(1 - a - n)} (2e^t)^{a + 2n}$$

for $n = -1, -2, \ldots$. Using Cauchy's theorem on residues and a well-known limit process, we obtain from (5.12) and the formulas above

$$\gamma(t, 0) = \left(\frac{2}{\pi}\right)^{1/2} \cot \pi a \left[\sum_{n=0}^{\infty} \frac{\Gamma(\frac{1}{2} + a + 2n)}{\Gamma(a + n)\Gamma(n + 1)} (2e^t)^{-(a + 2n)} \right.$$

$$\left. - \sum_{n=1}^{\infty} \frac{\Gamma(\frac{1}{2} - a + 2n)}{\Gamma(n)\Gamma(1 - a + n)} (2e^t)^{a - 2n} \right].$$

Due to the fact that $G_+(s) = G_-(-s)$, one obtains $\gamma(0, t) = \gamma(t, 0)$, $t \in R_+$. Once we have $\gamma(t, 0)$ and $\gamma(0, t)$, we can construct the resolvent kernel.

The efficiency of this method in solving integral equations depends, of course, on the needs of the application.

4.6   Perturbed Equations

If in Eq. (WH) we replace the free term $h(t)$ by a nonlinear term $h(t; x) = (hx)(t)$, where $x \to hx$ is a nonlinear operator, then we get the perturbed equation

$$x(t) = h(t; x) + \int_0^{\infty} k(t - s)x(s) \, ds, \qquad t \in R_+. \tag{6.1}$$

This section is devoted to the investigation of Eq. (6.1), under appropriate assumptions concerning the kernel $k(t)$ and the perturbing term $h(t; x)$. Let us mention that a similar problem concerning the Volterra integral equation was investigated in Section 2.11.

We shall give now a general result concerning Eq. (6.1).

Theorem 6.1   Consider Eq. (6.1) and assume that $k(t)$ satisfies conditions (2.1)–(2.3). Let E be any of the spaces $M(R_+, \mathscr{C})$, $L^p(R_+, \mathscr{C})$ $(1 \le p \le \infty)$, $C(R_+, \mathscr{C})$, $C_\ell(R_+, \mathscr{C})$, $C_0(R_+, \mathscr{C})$. If $x \to hx$ is a mapping from E into itself such that

$$|h(t; x) - h(t; y)|_E \le \lambda |x - y|_E \tag{6.2}$$

for $x, y \in E$, then there exists a unique solution $x \in E$ of Eq. (6.1), as long as λ is small enough.

Proof   The operator $x \to Tx = y$ from E into itself, where y is the solution of the equation

$$y(t) = h(t; x) + \int_0^\infty k(t - s)y(s)\, ds, \qquad t \in R_+, \tag{6.3}$$

is a contraction mapping. According to Theorem 2.1, $y \in E$ is uniquely determined by (6.3), under assumptions (2.1)–(2.3). Moreover, it can be expressed by means of the formula

$$y(t) = h(t; x) + \int_0^\infty \gamma(t, s)h(s; x)\, ds, \qquad t \in R_+. \tag{6.4}$$

If $k_0 \in L(R, R)$ is the majorant of $\gamma(t, s)$, i.e., $|\gamma(t, s)| \le k_0(t - s), 0 \le t, s < \infty$, then we obtain from (6.4)

$$|Tx_1 - Tx_2|_E \le (1 + |k_0|_L)\lambda |x_1 - x_2|_E \tag{6.5}$$

for any $x_1, x_2 \in E$. The inequality (6.5) shows that T is a contraction mapping. Theorem 6.1 is thus proved.

We shall consider below some applications of Theorem 6.1. Let us assume that the operator h is formally given by the formula

$$h(t; x) = h(t) + \int_0^\infty k_0(t, s)x(s)\, ds, \qquad t \in R_+, \tag{6.6}$$

with $h(t)$ and $k_0(t, s)$ satisfying appropriate conditions. Then Eq. (6.1) becomes

$$x(t) = h(t) + \int_0^\infty [k(t - s) + k_0(t, s)]x(s)\, ds \tag{6.7}$$

i.e., we obtain a linear equation with a perturbed kernel.

Corollary 1   Consider Eq. (6.7) and assume that $k(t)$ satisfies conditions (2.1)–(2.3). Let $k_0(t, s)$ be a measurable complex-valued function of (t, s) for $0 \le t, s < \infty$ such that there exists $k_0(t) \in L(R, R)$, with

$$|k_0(t, s)| \le k_0(t - s), 0 \le t, s < \infty.$$

If $h(t) \in M(R, \mathscr{C})$ and $|k_0|_L$ is small enough, then Eq. (6.7) has a unique solution $x(t) \in M$.

Indeed, $h(t; x)$ is an operator from M into itself and we have

$$|h(t; x) - h(t; y)|_M \leq |k_0|_L |x - y|_M$$

for any $x, y \in M$. Therefore, Theorem 6.1 applies without any difficulty.

Corollary 2 Let us consider Eq. (6.7) under assumptions (2.1)–(2.3). Assume further that $k_0(t, s)$ is a measurable complex-valued function such that

$$\operatorname*{ess\,sup}_{t \in R_+} \int_0^\infty |k_0(t, s)| \, ds \leq \lambda,$$

and

$$\operatorname*{ess\,sup}_{s \in R_+} \int_0^\infty |k_0(t, s)| \, dt \leq \lambda.$$

If E denotes any of the spaces $L^p(R_+, \mathscr{C})$, $1 \leq p \leq \infty$, then Eq. (6.7) has a unique solution $x(t) \in E$ for any $h \in E$, provided λ is small enough.

Theorem 2.8.2 assures that $h(t; x)$, given by (6.6), is an operator from E into itself. Moreover, the Lipschitz condition

$$|h(t; x) - h(t; y)|_E \leq \lambda |x - y|_E$$

holds for any $x, y \in E$.

Corollary 3 Assume that Eq. (6.7) satisfies the following conditions: (a) the kernel $k(t)$ is as described in Theorem 6.1; (b) the function $k_0(t, s)$ is measurable in (t, s), $0 \leq t, s < \infty$, and such that

$$\int_0^\infty \left[\int_0^\infty |k_0(t, s)|^p \, dt \right]^{q/p} ds \leq \lambda^q,$$

where $1 < p < \infty$ and $q = p/(p - 1)$, λ being a positive constant; (c) $h(t) \in L^p(R_+, \mathscr{C})$. Then there exists a unique solution $x(t) \in L^p(R_+, \mathscr{C})$ of (6.7), as long as λ is sufficiently small.

The proof of Corollary 3 follows also from Theorem 6.1. According to Theorem 2.8.3, the operator $x \to hx$ given by (6.6) is a completely continuous operator from L^p into itself such that its norm does not exceed λ.

A particular case of Corollary 3 corresponds to $p = q = 2$. Then, the condition concerning $k_0(t, s)$ takes the well-known form of square integrability

$$\int_0^\infty \int_0^\infty |k_0(t, s)|^2 \, dt \, ds \leq \lambda^2. \tag{6.8}$$

In other words, $k_0(t, s)$ must be square integrable and its L^2 norm has to be sufficiently small.

It is interesting to notice that Theorem 6.1 also applies when $x \to hx$ is the classical operator

$$h(t; x) = h(t, x(t)). \qquad (6.9)$$

For instance, in order to obtain an existence result for L^2 solutions, it suffices to assume that: (a) $h(t, x)$ is measurable in t, $t \in R_+$, for fixed x, and continuous in x, $x \in R$, for almost all $t \in R_+$; (b) there exists a positive constant λ such that $|h(t, x) - h(t, y)| \leq \lambda |x - y|$ for $x, y \in R$ and almost all $t \in R_+$; (c) $h(t, 0) \in L^2(R_+, R)$.

Corollary 4 If $k(t)$ satisfies (2.1)–(2.3) and conditions (a), (b), and (c) stated above, then the equation

$$x(t) = h(t, x(t)) + \int_0^\infty k(t - s)x(s)\, ds, \qquad t \in R_+ \qquad (6.10)$$

has a unique solution $x \in L^2$ for λ sufficiently small.

Indeed, from $|h(t, x(t))| \leq \lambda |x(t)| + |h(t, 0)|$, there results that $x \to hx$ is an operator from L^2 into itself. Inasmuch as

$$\int_0^\infty |h(t, x(t)) - h(t, y(t))|^2\, dt \leq \lambda^2 \int_0^\infty |x(t) - y(t)|^2\, dt,$$

we obtain

$$|h(t, x(t)) - h(t, y(t))|_{L^2} \leq \lambda |x - y|_{L^2}, \qquad (6.11)$$

i.e., the operator h satisfies a Lipschitz condition of the form (6.2).

Finally, we shall discuss one more example, concerned with the existence of the solutions in C_ℓ.

Corollary 5 Assume that $k(t)$ satisfies (2.1)–(2.3). Let $k_0(t, s)$ be a measurable function on $0 \leq t, s < \infty$ such that the following conditions are verified:

a. there exists $\lambda > 0$ with the property that

$$\int_0^\infty |k_0(t, s)|\, ds \leq \lambda, \qquad t \in R_+; \qquad (6.12)$$

b. to any $\varepsilon > 0$ and $t_0 \in R_+$, there corresponds $\delta > 0$ such that

$$\int_0^\infty |k_0(t, s) - k_0(t_0, s)|\, ds < \varepsilon \qquad \text{for} \quad |t - t_0| < \delta; \qquad (6.13)$$

c. for any A, $0 < A \leq \infty$, there exists

$$\lim_{t \to \infty} \int_0^A k_0(t, s)\, ds;$$

d. for any $\varepsilon > 0$ and $A > 0$, there exists $\delta > 0$ and $T > 0$ such that

$$\left| \int_E k_0(t, s)\, ds \right| < \varepsilon \tag{6.14}$$

as long as

$$E \subset [0, A), \qquad \text{mes } E < \delta, \qquad t > T. \tag{6.15}$$

Then Eq. (6.7) has a unique solution $x \in C_\ell$ whenever λ is sufficiently small.

The proof of Corollary 5 is a consequence of Theorem 6.1. We have to take into account the fact that the operator

$$(Kx)(t) = \int_0^\infty k_0(t, s)x(s)\, ds$$

acts from C_ℓ into itself and its norm does not exceed λ (see Dunford and Schwartz [1, Chapter IV, Section 13]).

Exercises

1. Consider the integral equation

$$x(t) = e^{i\xi t} + \int_0^\infty k(t - s)x(s)\, ds, \qquad t \in R_+,$$

with $k(t)$ satisfying conditions (2.1)–(2.3) and $\operatorname{Im} \xi \geq 0$. Prove that the solution is given by

$$x(t) = G_-(-\xi)\left(1 + \int_0^t \gamma(u, 0)e^{-i\xi u}\, du\right)e^{i\xi t}.$$

2. Consider the integral equation (WH) and assume that there exists $\tau \in R$ with the property $e^{-\tau t}|k(t)| \in L(R, R)$. If we denote

$$e^{-\tau t}k(t) = k_1(t), \qquad e^{-\tau t}x(t) = x_1(t), \qquad e^{-\tau t}h(t) = h_1(t),$$

then $x_1(t)$ satisfies the equation

(E) $$x_1(t) = h_1(t) + \int_0^\infty k_1(t - s)x_1(s)\, ds, \qquad t \in R_+.$$

Since $k_1(t) \in L(R, \mathscr{C})$, we can apply to the Eq. (E) the theory developed in this chapter. Formulate the main results of this chapter in the case of an Eq. (WH) whose kernel satisfies the above condition.

Hint: Observe that $\tilde{k}_1(s) = \tilde{k}(s + i\tau)$.

3. Discuss the Eq. (WH) whose kernel is given by

$$k(t) = \tfrac{1}{2}e^{-|t|}.$$

4. Discuss the Wiener–Hopf equation of the first kind

$$\int_0^\infty k(t-s)x(s)\,ds = h(t), \qquad t \in R_+.$$

5. Let $P(z) = \sum_{j=1}^{2m} a_j z^j$ and consider the integro-differential equation

$$P(d/dt)x(t) + \int_0^\infty k(t-s)x(s)\,ds = h(t), \qquad t \in R_+.$$

Apply the method of this chapter to the above equation.

6. Let $k(t) \in L(R, \mathscr{C})$ and consider the convolution operator K given by

$$(Kx)(t) = \int_0^\infty k(t-s)x(s)\,ds, \qquad t \in R_+.$$

If E denotes any of the spaces $L^p = L^p(R_+, \mathscr{C})$ $(1 \le p \le \infty)$, C, C_ℓ, C_0, then the spectrum of K on E consists of all complex numbers ζ such that $\zeta = \tilde{k}(s)$, $s \in R$, or $v(\zeta) = -\operatorname{ind}[\tilde{k}(s) - \zeta] \ne 0$.

Hint: If $\zeta \in \mathscr{C}$ is such that $\zeta - \tilde{k}(s) \ne 0$ for $s \in \bar{R}$, then

$$v(\zeta) = -\operatorname{ind}[\tilde{k}(s) - \zeta]$$

is well defined. Taking into account the results of this chapter and using the operator $\zeta^{-1}K$ instead of K, we get easily that any $\zeta \in \mathscr{C}$ with $v(\zeta) \ne 0$ belongs to the spectrum of K. There remains to discuss the case of those $\zeta \in C$ for which $\zeta = \tilde{k}(s)$, $s \in R$.

7. Consider the integral equation

$$x(t) = h(t) + \int_0^1 s^{-1}k(t/s)x(s)\,ds, \qquad t \in (0, 1).$$

Show that it can be reduced to an integral equation of the form (WH) by means of the change of variables $t = e^{-\tau}$, $s = e^{-\theta}$. The kernel of the new equation will be $k_1(\tau) = k(e^{-\tau})$. Formulate some results derived from the theory developed in this chapter.

8. Let $k(t)$ be a kernel satisfying condition (2.1) and consider on E the following operator

$$(Ax)(t) = x(t) - \int_0^\infty k(t-s)x(s)\,ds, \qquad t \in R_+.$$

Discuss the problem of existence of the inverse operator A^{-1} (in the algebra of continuous operators of E). Show that under conditions (2.2) and (2.3) there exists a unique inverse A^{-1}. If (2.2) holds, but $v = -\text{ind}[1 - \tilde{k}(s)] > 0$, then there exists infinitely many right inverse operators A^{-1}. What happens in the case $v < 0$?

Bibliographical Notes

The results of Sections 4.1–4.5 are due to M. G. Krein [1], excepting the remarks that some of them are valid when $E = M$ (see C. Corduneanu [11]). The results of Section 4.6 are due to the author [5].

As remarked by M. G. Krein [1], some results closely related to those given in this chapter were obtained earlier by Reisner [1] and Rapoport [1]. In particular, the significance of the index was recognized by Rapoport.

The exercises to Chapter IV are compiled following M. G. Krein's paper [1], the book by Gochberg and Feldman [1], and the compendium by Zabreiko *et al.*

In their joint paper [1], Gochberg and M. G. Krein investigated the vector case corresponding to Eq. (WH). The method developed in this chapter cannot be used in a straightforward manner in the vector case. By means of some general results on linear operators, Gochberg and M. G. Krein [1] have proved the main results for the vector equation (WH). Let us remark that condition (2.2) has to be replaced by $\det(I - \tilde{k}(s)) \neq 0$, $s \in R$. Instead of a single index, in the vector case there occurs a finite set of indices $v_1, v_2, \ldots, v_n$. They appear naturally in the factorization formula

$$[I - \tilde{k}(s)]^{-1} = G_+(s)Z(s)G_-(s),$$

where

$$Z(z) = \begin{pmatrix} z^{v_1} & 0 & 0 & \cdots & 0 \\ 0 & z^{v_2} & 0 & \cdots & 0 \\ \vdots & \vdots & \vdots & & \vdots \\ 0 & 0 & 0 & & z^{v_n} \end{pmatrix},$$

with $z = (s - i)/(s + i)$. Let us state a result concerning the vector equation (WH) with $\|k(t)\| \in L(R, R)$ and $\det(I - \tilde{k}(s)) \neq 0$ for $s \in R$. A necessary and sufficient condition that (WH) possess a unique solution $x \in E = E(R_+, R^n)$ for any $h \in E$ is that all the indices v_j, $j = 1, 2, \ldots, n$, be zero. If $v_j \geq 0$, $j = 1, 2, \ldots, n$, but at least one index is positive, then there exist infinitely many solutions. As usual, E stands for any one of the spaces

$$M, L^p \ (1 \leq p \leq \infty), C, C_\ell, C_0.$$

Various generalizations of the results given in this chapter have been obtained by Budjanu [1], Čebotarev [1], Čebotarev and Govorova [1], and

Čebotaru [1]. The last author gave a reduction method, i.e., a method that allows us to associate with an Eq. (WH) of arbitrary index an equation of the same form but whose index is zero.

The factorization problems we solved in Section 1.4 and applied in this chapter play—obviously—an important role in studying Eq. (WH). The key formula $[1 - \tilde{k}(s)]^{-1} = G_+(s)G_-(s)$, $s \in R$, is a special case of the following equation:

(H) $$A(s)H_+(s) + B(s)H_-(s) = C(s),$$

with $H_+(s) = G_+(s)$ and $H_-(s) = 1/G_-(s)$. The functions $H_+(s)$ and $H_-(s)$ enjoy the same properties as $G_+(s)$ and $G_-(s)$, respectively. But, finding the functions $H_+(s)$ (holomorphic for $\text{Im } s > 0$ and continuous for $s \in R$) and $H_-(s)$ (holomorphic for $\text{Im } s < 0$ and continuous for $s \in R$) such that (H) holds for any $s \in R$ represents what is usually called a Hilbert problem. Consequently, the investigation of Eq. (WH) can be reduced to the discussion of a Hilbert problem. This is a very efficient method even for practical purposes (i.e., to solve a given integral equation of the form (WH) or the corresponding equation of the first kind). For some examples on this subject we refer the reader to the compendium by Zabreiko *et al.* Actually, Hilbert's problem covers a larger area than solving integral equations, and for a better understanding of this fact we refer the reader to the well-known book by Noble [1]. The book by Muskhelishvili [1] is another important reference. In this book, mainly devoted to the theory of singular integral equations of the form

(C) $$A(t)\varphi(t) + (B(t)/\pi i) \int_\Gamma [\varphi(s) \, ds/(s - t)] = c(t),$$

with Cauchy's integral along a certain curve Γ, the connection with the Hilbert problem is largely emphasized. There are many resemblances between the theory of Eq. (WH) as it was developed in this chapter, and the theory of equations of the form (C).

We should point out that the reduction of various problems to the solution of a Hilbert problem is commonly called the Wiener–Hopf technique. Historically, a very good reference is Wiener and Hopf [1].

Before concluding these notes, we want to call the reader's attention to the paper by Gochberg [1] and to the already quoted book by Gochberg and Feldman [1]. A short presentation of the main ideas of the first paper appears in the introductory part of the book. Constructing a very general theory, mainly based on ring theory, it is possible to give a unified treatment of Eq. (WH), of the singular integral equation (C), and of a discrete analogue of (WH). Very likely, it is also possible to cover other classical areas. Let us mention now a result from Gochberg and Feldman [1]: Consider the scalar equation (WH) and assume that $k(t)$ satisfies (2.1). Then conditions (2.2) and (2.3) are necessary and sufficient in order that the "approximate" equation

$$x_T(t) = h(t) + \int_0^T k(t - s)x_T(s)\, ds$$

possess a unique solution for $T \geq T_0$, with T_0 sufficiently large, such that $|x - x_T|_{L^p} \to 0$ as $T \to \infty$, where x denotes the solution of Eq. (WH) for arbitrary $h \in L^p$. It is assumed that $x_T(t) = 0$ for $t > T$. Particular attention is paid by these authors to the theory of difference-integral operators of the form

$$(Ax)(t) = \sum_{j=0}^{\infty} a_j x(t - t_j) + \int_0^{\infty} k(t - s)x(s)\, ds$$

with $\Sigma\, |a_j| < +\infty$ and $k \in L(R, \mathscr{C})$.

Further Methods and Topics

This last chapter of the book is aimed at illustrating several methods and topics of the theory of nonlinear integral or integro-differential equations, and at indicating an interesting application for some results we shall establish below to the dynamics of nuclear reactors. Such methods and topics are often encountered in the research literature, and they did not find place in the preceding chapters.

We shall discuss and apply the so-called "energy method," which can be compared with Liapunov's function method in the theory of ordinary differential equations. The origin of this method in the field of integro-differential equations can be traced to Volterra's work (see, for instance, Volterra [1]).

The concept of positive-definite kernel and some of its applications to the investigation of the behavior of solutions will also be emphasized.

Another powerful tool in studying the asymptotic behavior of the solutions of integral or integro-differential equations is furnished by the application of tauberian theorems. Some of the results we shall include in this chapter rely upon tauberian arguments.

Finally, the applications of the general results to the dynamics of nuclear reactors seem to be a strongly stimulating reason for the research activity in this field during the last ten–twelve years.

5.1 Some Results of Levin

Let us consider the integro-differential equation

$$\dot{x}(t) = - \int_0^t k(t, s)\varphi(x(s))\, ds - b(t) \tag{1.1}$$

on the positive half-axis, and assume that all the quantities occurring above are scalar and real.

The most general assumptions under which the "energy method" works and leads to very useful results have been indicated by Levin [3]. We shall closely follow his presentation of general results, as well as some applications to particular cases.

Before formulating the results, we shall state the hypotheses.

H_1:   The kernel $k(t, s)$ is continuous on the set

$$D = \{(t, s) : 0 \leq s \leq t < +\infty\}, \tag{1.2}$$

is continuously differentiable of the third order on the subset

$$\hat{D} = \{(t, s) : 0 \leq s < t < \infty\}, \tag{1.3}$$

and satisfies the inequalities

$$k(t, s) \geq 0, \qquad k_t(t, s) \leq 0, \qquad k_{tt}(t, s) \geq 0,$$
$$k_s(t, s) \geq 0, \qquad k_{ts}(t, s) \leq 0, \qquad k_{tts}(t, s) \geq 0.$$

H_2:   For every function $\theta(t) \in C(R_+, R)$, the following formulas hold:

$$(d/dt) \int_0^t k(t, s)\theta(s)\, ds = k(t, t)\theta(t) + \int_0^t k_t(t, s)\theta(s)\, ds,$$

$$(d/dt) \int_0^t k_s(t, s) \left[\int_s^t \theta(u)\, du \right]^2 ds = \int_0^t k_{ts}(t, s) \left[\int_s^t \theta(u)\, du \right]^2 ds$$

$$- 2\theta(t)k(t, 0) \int_0^t \theta(s)\, ds + 2\theta(t) \int_0^t k(t, s)\theta(s)\, ds,$$

and a similar formula with $k_{ts}(t, s)$ instead of $k_s(t, s)$.

Remark   The formulas occurring in H_2 are not generally verified under assumption H_1, since we required the smoothness of $k(t, s)$ in $\hat{D}$ only.

H_3:   The following inequalities hold:

$$\sup_{0 \le t < \infty} k(t, t) < +\infty,$$

$$\inf_{0 \le t < \infty} \int_0^t k_t(t, s) \, ds > -\infty,$$

$$\limsup_{t \to \infty} \int_{t-\delta}^t (t - s)^2 k_{ts}(t, s) \, ds < 0,$$

for every $\delta > 0$.

Remark   The kernel $k(t, s) = \exp\{-(t - s) \cdot (t + 2)^2\}$ satisfies H_1, H_2, and the first two inequalities in H_3. However, it does not satisfy the last condition in H_3.

H_4: $\varphi(x)$ is a continuous function from R into itself such that $x\varphi(x) > 0$ for $x \ne 0$ and

$$\phi(x) = \int_0^x \varphi(\xi) \, d\xi \to \infty \qquad \text{as} \quad |x| \to \infty.$$

H_5: $b(t) \in C(R_+, R)$ is continuously differentiable of the second order for $t > 0$, and there exists a function $c(t)$ with the same properties as $b(t)$ such that

$$[b^{(j)}(t)]^2 \le c^{(j)}(t)(\partial^j/\partial t^j)k(t, 0), \qquad j = 0, 1, 2, \tag{1.4}$$

for any $t > 0$.

Remark   From the first three inequalities in H_1, with $s = 0$, we obtain $k(t, 0) \ge 0$, $k_t(t, 0) \le 0$, $k_{tt}(t, 0) \ge 0$. These imply the boundedness of $k_t(t, 0)$ on any half-axis $[t_0, \infty)$, $t_0 > 0$. From (1.4) there results $c(t) \ge 0$, $\dot{c}(t) \le 0$, $\ddot{c}(t) \ge 0$, from which we get the boundedness of $\dot{c}(t)$. Therefore, our assumptions imply that $\dot{b}(t)$ is bounded on any half-axis $[t_0, \infty)$, $t_0 > 0$.

Theorem 1.1   Assume that hypotheses H_1–H_5 are satisfied. Then for each $x_0 \in R$, there exists at least one solution $x(t)$ of (1.1) defined on R_+ such that $x(0) = x_0$. Moreover,

$$\lim_{t \to \infty} x^{(j)}(t) = 0, \qquad j = 0, 1, 2. \tag{1.5}$$

Proof   Equation (1.1), with the initial condition $x(0) = x_0$, is equivalent to the integral equation

$$x(t) = x_0 - \int_0^t b(s) \, ds - \int_0^t du \int_0^u k(u, s)\varphi(x(s)) \, ds.$$

Changing the order in the double integral we obtain the following integral equation:

$$x(t) = x_0 - \int_0^t b(s)\,ds - \int_0^t \left\{ \int_s^t k(u, s)\,du \right\} \varphi(x(s))\,ds. \tag{1.6}$$

Now it is possible to apply a theorem of local existence (see, for instance, C. Corduneanu [12], Theorem 6.2) to (1.6). There results the existence of at least one solution $x(t)$, defined on a certain interval $[0, T]$, $T > 0$.

Let us now consider for $t \in [0, T]$ the "energy function"

$$E(t) = \phi(x(t)) + \tfrac{1}{2}k(t, 0)\left(\int_0^t \varphi(x(s))\,ds \right)^2 + b(t) \int_0^t \varphi(x(s))\,ds$$

$$+ \tfrac{1}{2}c(t) + \tfrac{1}{2} \int_0^t k_s(t, s)\left(\int_s^t \varphi(x(u))\,du \right)^2 ds. \tag{1.7}$$

It is trivial to check that the first and fifth terms in the second member of (1.7) are nonnegative.

Concerning the sum

$$\tfrac{1}{2}k(t, 0)\left(\int_0^t \varphi(x(s))\,ds \right)^2 + b(t) \int_0^t \varphi(x(s))\,ds + \tfrac{1}{2}c(t),$$

which is a quadratic polynomial in $\int_0^t \varphi(x(s))\,ds$, we can also assert that it is nonnegative because of the inequality in H_5 with $j = 0$. Therefore, we can state

$$E(t) \geq 0, \qquad t \in [0, T]. \tag{1.8}$$

We now differentiate $E(t)$, taking into account (1.1) and H_3. There results

$$\dot{E}(t) = \tfrac{1}{2}k_t(t, 0)\left(\int_0^t \varphi(x(s))\,ds \right)^2 + b(t) \int_0^t \varphi(x(s))\,ds$$

$$+ \tfrac{1}{2}\dot{c}(t) + \tfrac{1}{2} \int_0^t k_{ts}(t, s)\left(\int_s^t \varphi(x(u))\,du \right)^2 ds.$$

Similar arguments to those used above lead to

$$\dot{E}(t) \leq 0, \qquad t \in [0, T]. \tag{1.9}$$

The inequalities (1.8) and (1.9) obviously give on $[0, T]$:

$$\phi(x(t)) \leq E(t) \leq E(0) = \phi(x_0) + \tfrac{1}{2}c(0). \tag{1.10}$$

Hence, there exists an a priori estimate for the solutions of (1.1), namely,

$$|x(t)| \leq K(x_0), \qquad t \in [0, T], \tag{1.11}$$

with $K(x_0) = \phi_0^{-1}(\phi(x_0) + \tfrac{1}{2}c(0))$, where $\phi_0(u)$ is an increasing function for $u \geq 0$ such that $\phi_0(u) \leq \phi(\pm u)$.

Using well-known arguments, one can prove that $x(t)$ exists on R_+ (more precisely, it can be extended on R_+) and

$$|x(t)| \leq K(x_0), \qquad t \in R_+. \tag{1.12}$$

Let us now differentiate both sides of Eq. (1.1), taking into account H_2. We obtain

$$\dddot{x}(t) = -k(t, t)\varphi(x(t)) - \int_0^t k_t(t, s)\varphi(x(s))\, ds - \dot{b}(t).$$

This inequality, together with (1.12) and our hypotheses (see H_3 and H_5), yields

$$\limsup_{t \to \infty} |\dddot{x}(t)| < +\infty. \tag{1.13}$$

We claim that (1.12) and (1.13) imply

$$\limsup_{t \to \infty} |\dot{x}(t)| < +\infty. \tag{1.14}$$

Indeed, let us denote

$$\ddot{x}(t) - x(t) = h(t), \qquad t \geq t_0 > 0. \tag{1.15}$$

It follows that $h(t) \in C([t_0, \infty), R)$, i.e., $h(t)$ is a continuous bounded function on $[t_0, \infty)$. Since we know that $x(t)$ is bounded on R_+, there results from (1.15) that

$$x(t) = ae^{-t} - \tfrac{1}{2}\left[e^t \int_t^\infty e^{-s}h(s)\, ds + e^{-t}\int_{t_0}^t e^s h(s)\, ds \right],$$

where a is an arbitrary constant. From the above formula for $x(t)$, one easily obtains (1.14).

Let us now differentiate $\dot{E}(t)$, taking into account the expression we found above for $\dddot{x}(t)$ and H_2. There results

$$\ddot{E}(t) = \tfrac{1}{2}k_{tt}(t, 0)\left(\int_0^t \varphi(x(s))\, ds \right)^2 + \dot{b}(t)\int_0^t \varphi(x(s))\, ds + \tfrac{1}{2}\ddot{c}(t)$$

$$+ \tfrac{1}{2}\int_0^t k_{tts}(t, s)\left(\int_s^t \varphi(x(u))\, du \right)^2 ds - \varphi(x(t))[\ddot{x}(t) + k(t, t)\varphi(x(t))].$$

Considering H_1 and H_5, we get

$$\ddot{E}(t) \geq -\varphi(x(t))[\ddot{x}(t) + k(t, t)\varphi(x(t))].$$

Taking into account (1.12), (1.13), and H_3, it follows that $\ddot{E}(t)$ is bounded below on $[t_0, \infty)$. To summarize, $E(t) \geq 0$, $\dot{E}(t) \leq 0$, and $\ddot{E}(t) \geq -K_0 > -\infty$ on $[t_0, \infty)$.

We now show that

$$\lim_{t \to \infty} \dot{E}(t) = 0. \tag{1.16}$$

Indeed, if we assume that (1.16) does not hold, there exist a $\lambda > 0$ and a sequence $\{t_n\}$, with $t_n \to \infty$ as $n \to \infty$, such that $\dot{E}(t_n) \leq -\lambda < 0$. Consider the intervals $I_n = [t_n - \lambda/2K_0, t_n]$, and take $n \geq N$ such that $t_n > \lambda/2K_0$. Since for $t \in I_n$ we have

$$\dot{E}(t) = \dot{E}(t_n) + \ddot{E}(\theta_n)(t - t_n) \leq -\lambda + (\lambda/2) = -(\lambda/2),$$

where $t < \theta_n < t_n$, there results

$$E(t_n - (\lambda/2K_0)) - E(t_n) \geq (\lambda/2)(\lambda/2K_0) = \lambda^2/4K_0,$$

which constitutes a contradiction of the fact that $E(t)$ decreases to a finite limit as $t \to \infty$. Consequently, (1.16) is proven.

From (1.16) and from the formula for $\dot{E}(t)$, we easily obtain (see H_1 and H_5)

$$\lim_{t \to \infty} \int_0^t k_{ts}(t, s) \left(\int_s^t \varphi(x(u)) \, du \right)^2 ds = 0. \tag{1.17}$$

If we assume now that $x(t)$ does not vanish at infinity, from (1.12), (1.14), H_1, and H_4 we derive the following property: there exist positive constants λ, δ, and a sequence $\{t_n\}$, with $t_n \to \infty$ as $n \to \infty$, such that

$$\int_{t_n - \delta}^{t_n} k_{ts}(t_n, s) \left(\int_s^{t_n} \varphi(x(u)) \, du \right)^2 ds \leq \lambda \int_{t_n - \delta}^{t_n} k_{ts}(t_n, s)(t_n - s)^2 \, ds \leq 0.$$

Taking into account (1.17), there results

$$\lim_{n \to \infty} \int_{t_n - \delta}^{t_n} (t_n - s)^2 k_{ts}(t, s) \, ds = 0,$$

which contradicts the last condition in H_3. Therefore, $\lim x(t) = 0$ as $t \to \infty$. In other words, we have obtained (1.5) for $j = 0$.

In order to prove (1.5) for $j = 1$, we can use an argument quite similar to that used in the proof of (1.16). We invite the reader to accomplish the proof.

It remains to prove that (1.5) also holds for $j = 2$. From the formula

$$\ddot{x}(t) = -k(t, t)\varphi(x(t)) - \int_0^t k_t(t, s)\varphi(x(s)) \, ds - \dot{b}(t),$$

one finds easily that the first and third terms go to zero as $t \to \infty$. Indeed, according to H_3, $k(t, t)$ is bounded. As seen above, $x(t) \to 0$ as $t \to \infty$. Since $\varphi(0) = 0$, it follows that the first term vanishes at infinity. From the first three inequalities in H_1, with $s = 0$, we get $\lim(\partial/\partial t)k(t, 0) = 0$ as $t \to \infty$, by the same arguments we used in proving (1.16). Taking into account H_5, we obtain

$\lim \dot{b}(t) = 0$ as $t \to \infty$. Concerning the middle term in the formula giving $\ddot{x}(t)$, we shall write it in the form

$$\int_0^t k_t(t, s)\varphi(x(s))\,ds = \int_0^T k_t(t, s)\varphi(x(s))\,ds + \int_T^t k_t(t, s)\varphi(x(s))\,ds. \qquad (1.18)$$

For any $\varepsilon > 0$, according to H_1, H_3, and (1.5) for $j = 0$, there exists $T = T(\varepsilon) > 0$ such that

$$\left| \int_T^t k_t(t, s)\varphi(x(s))\,ds \right| < \varepsilon,$$

whenever $t \geq T$. Since the first three inequalities in H_1 imply $k_t(t, s) \to 0$ as $t \to \infty$ for any fixed s and $k_{ts}(t, s) \leq 0$, one can write

$$k_t(t, T) \leq k_t(t, s) \leq k_t(t, 0), \qquad 0 \leq s \leq T,$$

which means that $k_t(t, s) \to 0$ as $t \to \infty$, uniformly on $0 \leq s \leq T$. Hence, the first term in the right member of (1.18) tends to zero as $t \to \infty$. Consequently,

$$\lim_{t \to \infty} \int_0^t k_t(t, s)\varphi(x(s))\,ds = 0,$$

from which we get (1.5) with $j = 2$. Thus, Theorem 1.1 is proven.

Remark The uniqueness of the solution can be proved by elementary arguments (for instance, using Gronwall's inequality) if we assume that $\varphi(x)$ satisfies a local Lipschitz condition.

We shall now investigate the equation

$$\dot{x}(t) = -\int_0^t k(t, s)\varphi(x(s))\,ds - b(t) + f(t), \qquad (1.19)$$

where $f(t)$ is subject to the following condition.

H_6: $f(t) \in C(R_+, R)$, is continuously differentiable for $t > 0$, and its derivative is such that

$$\dot{f}(t) \in L(R_+, R), \qquad \sup_{0 \leq t < \infty} |\dot{f}(t)| < +\infty. \qquad (1.20)$$

Theorem 1.2 Consider Eq. (1.19) and assume that hypotheses H_1–H_6 hold. Moreover, let $\varphi(x)$ be continuously differentiable such that

$$\limsup_{|x| \to \infty} [\,|\varphi(x)|/\phi(x)] < \infty \qquad (1.21)$$

be verified. Then any solution of (1.19) satisfies

$$\lim_{t \to \infty} x^{(j)}(t) = 0, \qquad j = 0, 1. \qquad (1.22)$$

Proof   We denote

$$F(t) = \int_0^t |f(s)|\, ds, \qquad t \in R_+ . \tag{1.23}$$

According to H_4 and assumption (1.21), there exists a positive constant A such that

$$|\varphi(x)| \le A[1 + \phi(x)], \qquad x \in R. \tag{1.24}$$

Let $x_0 \in R$ be given and denote by $x(t)$ a solution of (1.19) with $x(0) = x_0$. Such a solution exists on a certain interval $[0, T]$, $T > 0$. As in Theorem 1.1 we shall find an estimate for $|x(t)|$ that is independent of T. This implies, as seen above, the existence of the solution on the entire half-axis.

Consider now the function

$$V(t) = [1 + E(t)] \exp\{-AF(t)\} \ge 0,$$

with $E(t)$ given by (1.7). By differentiating $V(t)$ we obtain after some calculation

$$\dot{V}(t) = -A|f(t)|V(t) + \Big\{ \varphi(x(t))f(t)$$

$$+ \tfrac{1}{2}k_t(t,0)\Big(\int_0^t \varphi(x(s))\, ds\Big)^2 + b(t)\int_0^t \varphi(x(s))\, ds + \tfrac{1}{2}\dot{c}(t)$$

$$+ \tfrac{1}{2}\int_0^t k_{ts}(t,s)\Big(\int_s^t \varphi(x(u))\, du\Big)^2 ds \Big\} \exp\{-AF(t)\}.$$

Taking into account our assumptions we can write

$$\dot{V}(t) \le -A|f(t)|V(t) + \varphi(x(t))f(t)\exp\{-AF(t)\}$$

and, furthermore,

$$\dot{V}(t) \le -\{A[1 + \phi(x(t))] - |\varphi(x(t))|\}|f(t)| \exp\{-AF(t)\}.$$

The last inequality and (1.24) imply $\dot{V}(t) \le 0$ for all t for which $x(t)$ is defined. Consequently, we can write the inequality

$$\phi(x(t))\exp\{-AF(t)\} \le V(t) \le V(0) = 1 + \phi(x_0) + \tfrac{1}{2}c(0).$$

We now find

$$\phi(x(t)) \le [1 + \phi(x_0) + \tfrac{1}{2}c(0)] \exp\Big\{A \int_0^\infty |f(t)|\, dt\Big\},$$

which constitutes the a priori estimate for $x(t)$. It shows that $x(t)$ can be extended on the half-axis R_+ and moreover, considering H_4, that it is bounded on R_+:

$$\sup_{0 \le t < \infty} |x(t)| < +\infty. \tag{1.25}$$

Let us now differentiate both sides of Eq. (1.19):

$$\ddot{x}(t) = -k(t, t)\varphi(x(t)) - \int_0^t k_t(t, s)\varphi(x(s))\, ds - \dot{b}(t) + \dot{f}(t).$$

According to (1.25) and our assumptions (see mainly H_3, H_5, and H_6), we can state

$$\limsup_{t \to \infty} |\ddot{x}(t)| < +\infty. \tag{1.26}$$

Following the same argument as in the proof of Theorem 1.1, we obtain from (1.25) and (1.26)

$$\limsup_{t \to \infty} |\dot{x}(t)| < +\infty. \tag{1.27}$$

The next step consists in finding an inequality of the form

$$D\dot{V}(t) \ge -L > -\infty, \tag{1.28}$$

where D stands for the right-hand derivative. The necessity of introducing this generalized derivative is justified by the fact that $(d/dt)|f(t)|$ may not exist, while $D|f(t)|$ exists at any point and $|D|f(t)|| = |\dot{f}(t)|$. The following (quite intricate) expression is found for $D\dot{V}(t)$:

$$\begin{aligned}
D\dot{V}(t) = {}& -A|f(t)|\dot{V}(t) - AV(t)D|f(t)| \\
&+ \left[\left\{\tfrac{1}{2}k_{tt}(t, 0)\left(\int_0^t \varphi(x(s))\, ds\right)^2 + \dot{b}(t)\int_0^t \varphi(x(s))\, ds + \tfrac{1}{2}\ddot{c}(t)\right\}\right. \\
&- A|f(t)|\left\{\tfrac{1}{2}k_t(t,0)\left(\int_0^t \varphi(x(s))\, ds\right)^2 + b(t)\int_0^t \varphi(x(s))\, ds + \tfrac{1}{2}\dot{c}(t)\right\} \\
&+ \tfrac{1}{2}\left\{-A|f(t)|\int_0^t k_{ts}(t,s)\left(\int_s^t \varphi(x(u))\, du\right)^2 ds\right. \\
&\left.+ \int_0^t k_{tts}(t,s)\left(\int_s^t \varphi(x(u))\, du\right) ds\right\} \\
&+ \{\varphi'(x(t))\dot{x}(t)f(t) + \varphi(x(t))[-A|f(t)|f(t) \\
&\left.+ 2\dot{f}(t) - \ddot{x}(t) - k(t,t)\varphi(x(t))]\}\right] \exp\{-AF(t)\}.
\end{aligned}$$

In order to obtain the expression above after directly calculating $D\dot{V}(t)$, it is necessary to take into account the equation for $\ddot{x}(t)$ and to integrate by parts, which yields

$$\int_0^t k_{ts}(t, s)\left(\int_s^t \varphi(x(u))\, du\right) ds = k_t(t, 0)\int_0^t \varphi(x(u))\, du + \int_0^t k_t(t, s)\varphi(x(s))\, ds.$$

It is now very easy to obtain for $D\dot{V}(t)$ an estimate of the form (1.28). Indeed, taking into account that $V(t) \geq 0$, $\dot{V}(t) \leq 0$, the boundedness of $x(t)$, $\dot{x}(t)$, and $\ddot{x}(t)$, as well as our assumptions, it follows that $-A|f(t)|\dot{V}(t) \geq 0$, $AV(t)D|\dot{f}(t)|$ is bounded in absolute value by $AV(0)\sup|\dot{f}(t)|$, $0 < t < \infty$, and the first three expressions in curly brackets give nonnegative quantities, while the fourth one is bounded. Since $\exp\{-AF(t)\}$ is also bounded on R_+, the inequality (1.28) is established.

From $V(t) \geq 0$, $\dot{V}(t) \leq 0$, and (1.28), one obtains (see the proof of Theorem 1.1)

$$\lim_{t \to \infty} \dot{V}(t) = 0. \tag{1.29}$$

If we consider the formula giving $\dot{V}(t)$ and notice that H_6 implies $\lim f(t) = 0$ as $t \to \infty$ (see Barbălat's lemma in the proof of Theorem 2.2), one obtains

$$\lim_{t \to \infty} \int_0^t k_{ts}(t, s)\left(\int_s^t \varphi(x(u))\, du\right)^2 ds = 0.$$

The above formula is nothing but (1.17) and we get, as seen before, $\lim x(t) = 0$ as $t \to \infty$. (1.22) is thus proved for $j = 0$. It now follows easily that $\lim \dot{x}(t) = 0$ as $t \to \infty$, using elementary arguments (the mean-value theorem).

Remark Generally, (1.22) does not hold under our assumptions for $j = 2$. This situation occurs because $\dot{f}(t)$ does not necessarily vanish at infinity (as $\dot{b}(t)$ does).

Before concluding this section, we shall remark that the "energy method" used above also allows us to investigate integro-differential systems of the form

$$\dot{x}(t) = -\int_0^t k(t, s)\varphi(x(s))\, ds + f(t, x(t)),$$

under convenient assumptions with respect to $f(t, x)$. Corresponding results can be found in the paper by Levin [3].

5.2 A System Occurring in the Dynamics of Nuclear Reactors

We shall now consider the system of integro-differential equations

$$\dot{u}(t) = -\int_{-\infty}^{\infty} \alpha(x)T(x, t)\, dx,$$

$$T_t(x, t) = T_{xx}(x, t) + \eta(x)\varphi(u(t)), \tag{2.1}$$

with the initial conditions

$$u(0) = u_0, \qquad T(x, 0) = f(x), \qquad x \in R. \tag{2.2}$$

The prescribed functions $\alpha(x)$, $\eta(x)$, $f(x)$, and $\varphi(u)$ are real-valued, $u_0 \in R$ is also given, and $u(t)$ ($t \in R_+$) and $T(x, t)$ ($x \in R$, $t \in R_+$) are the unknowns.

In the particular case $\varphi(u) = -1 + \exp u$, the system (2.1) may be viewed as a mathematical model describing the behavior of a continuous-medium nuclear reactor. The unknowns quantities $u(t)$ and $T(x, t)$ stand for the deviations of the logarithm of the total reactor power and temperature from their equilibrium values. The reactor is modeled as a doubly infinite rod, a feature that allows us to simplify the treatment of problems. A more realistic assumption would require us to replace the infinite interval of integration in (2.1) by a finite one and to impose appropriate boundary conditions on $T(x, t)$.

Before formulating the hypotheses and results, let us make some formal considerations that show the way of reducing the system (2.1) to an integro-differential equation of the form

$$\dot{u}(t) = -\int_0^t k(t - s)\varphi(u(s))\, ds - b(t). \tag{2.3}$$

It is well known that under convenient conditions, the solution of the heat equation $T_t(t, x) = T_{xx}(x, t) + F(x, t)$, with the initial condition $T(x, 0) = f(x)$, $x \in R$, is given for $t > 0$ by

$$T(x, t) = \int_{-\infty}^{\infty} G(x - \xi, t)f(\xi)\, d\xi + \int_0^t \int_{-\infty}^{\infty} G(x - \xi, t - s)F(\xi, s)\, d\xi\, ds,$$

where

$$G(x, t) = (4\pi t)^{-1/2} \exp\{-(x^2/4t\}. \tag{2.4}$$

If we assume for the moment that $u(t)$ is known, then the second equation in (2.1) together with the second initial condition (2.2) yield

$$T(x, t) = \int_{-\infty}^{\infty} G(x - \xi, t)f(\xi)\, d\xi$$
$$+ \int_0^t \int_{-\infty}^{\infty} G(x - \xi, t - s)\eta(\xi)\varphi(u(s))\, d\xi\, ds. \tag{2.5}$$

Now we substitute $T(x, t)$ given by (2.5) in the first equation (2.1). The result is an equation of the form (2.3) for $u(t)$, with

$$k(t) = \int_{-\infty}^{\infty} \alpha(x)\, dx \int_{-\infty}^{\infty} G(x - \xi, t)\eta(\xi)\, d\xi, \tag{2.6}$$

$$b(t) = \int_{-\infty}^{\infty} \alpha(x)\, dx \int_{-\infty}^{\infty} G(x - \xi, t)f(\xi)\, d\xi. \tag{2.7}$$

In order to get a more convenient form for $k(t)$ and $b(t)$, we shall use Parseval's formula for Fourier transforms. Of course, we assume that all the operations we shall perform are valid. Later on, conditions that assure this validity will be formulated.

First of all, let us notice that the Fourier transform of $G(x, t)$, regarded as a function of x, is given by

$$\tilde{G}(\xi, t) = \exp\{-\xi^2 t\}, \tag{2.8}$$

for any $t > 0$ (see, for instance, Bateman and Erdelyi: "Tables of Integral Transforms"; McGraw-Hill, 1954).

If we denote

$$\rho(x) = \int_{-\infty}^{\infty} G(x - \xi, t)\eta(\xi)\,d\xi, \tag{2.9}$$

then (2.6) becomes

$$k(t) = \int_{-\infty}^{\infty} \alpha(x)\rho(x)\,dx. \tag{2.10}$$

Parseval's formula applied to (2.10) yields

$$k(t) = (2\pi)^{-1}\int_{-\infty}^{\infty} \overline{\tilde{\alpha}(x)}\tilde{\rho}(x)\,dx. \tag{2.11}$$

Since $\alpha(x)$ is real-valued, there results $\overline{\tilde{\alpha}(x)} = \tilde{\alpha}(-x)$. We get further

$$k(t) = \pi^{-1}\int_{0}^{\infty} \mathrm{Re}\{\tilde{\alpha}(-x)\tilde{\rho}(x)\}\,dx, \tag{2.12}$$

if we consider that a formula similar to (2.11) holds with $\tilde{\alpha}(x)\tilde{\rho}(x)$ as integrand.

But (2.8) and (2.9) show that

$$\tilde{\rho}(x) = \tilde{\eta}(x)\exp\{-x^2 t\}, \tag{2.13}$$

for any $t > 0$. Therefore, (2.12) and (2.13) lead to

$$k(t) = \pi^{-1}\int_{0}^{\infty} h_1(x)\exp\{-x^2 t\}\,dx, \tag{2.14}$$

where

$$h_1(x) = \mathrm{Re}\{\tilde{\alpha}(-x)\tilde{\eta}(x)\}. \tag{2.15}$$

Starting now from (2.7) instead of (2.6), we obtain in the same way

$$b(t) = \pi^{-1}\int_{0}^{\infty} h_2(x)\exp\{-x^2 t\}\,dx, \tag{2.16}$$

with

$$h_2(x) = \mathrm{Re}\{\tilde{\alpha}(-x)\tilde{f}(x)\}. \tag{2.17}$$

The formulas (2.14)–(2.17) will play a fundamental role in the approach we shall follow in studying the system (2.1).

Let us now state a lemma that constitutes a basic tool in the proof of Theorem 2.1—the main result of this section.

Lemma 2.1   Consider Eq. (2.3) and assume that H_4 holds for φ. Let us assume further that:

1. $k(t) \in C(R_+, R)$, $k(t) \not\equiv k(0)$, and $(-1)^j k^{(j)}(t) \geq 0$ on $0 < t < \infty$, $j = 0$, 1, 2, 3;

2. $b(t) \in C(R_+, R)$ and is continuously differentiable of the second order for $t > 0$;

3. there exists a function $c(t)$ with the same properties as $b(t)$ and such that

$$[b^{(j)}(t)]^2 \leq k^{(j)}(t)c^{(j)}(t), \qquad j = 0, 1, 2. \tag{2.18}$$

Then for each $u_0 \in R$ there exists a solution $u(t)$ of (2.3) on R_+ such that $u(0) = u_0$ and

$$\lim_{t \to \infty} u^{(j)}(t) = 0, \qquad j = 0, 1, 2. \tag{2.19}$$

Proof   The proof follows straightforwardly from Theorem 1.1. Our assumptions obviously imply those of Theorem 1.1. Some elementary calculation is needed in order to check the validity of H_3.

A direct proof can be obtained by using the function

$$E(t) = \phi(x(t)) + \tfrac{1}{2}k(t)\left(\int_0^t \varphi(x(s))\,ds\right)^2 + b(t)\int_0^t \varphi(x(s))\,ds$$
$$+ \tfrac{1}{2}c(t) - \tfrac{1}{2}\int_0^t k(t-s)\left(\int_s^t \varphi(x(u))\,du\right)^2 ds.$$

Let us now formulate the assumptions under which the existence and behavior of solutions for (2.1) and (2.2) can be investigated by using Lemma 2.1.

A_1:   The functions $\alpha(x)$ and $\eta(x)$ are measurable and such that

$$\alpha(x), \eta(x) \in L^2(R, R). \tag{2.20}$$

Moreover, $\eta(x)$ is locally Hölder continuous.

As a first consequence of A_1, we mention the fact that both $\tilde{\alpha}(x)$ and $\tilde{\eta}(x)$ are defined as generalized Fourier transforms (see Exercise 4, Section 1.3) and belong to L^2. This obviously implies

$$h_1(x) \in L(R, R). \tag{2.21}$$

Our next assumption is

$$A_2: \qquad h_1(x) \geq 0, \int_0^\infty h_1(x)\, dx > 0.$$

With respect to $f(x)$, we assume that:

$$A_3: \qquad f(x) \in L^2(R, R) \cap C(R, R).$$

From A_3 there results that $\tilde{f}(x)$ is also defined and belongs to L^2, and $h_2(x)$ given by (1.17) satisfies

$$h_2(x) \in L(R, R). \tag{2.22}$$

Finally, one more hypothesis is needed in order to be able to state the main result of this section.

$A_4:$ There exist a measurable function $h_3(x)$ with $h_3(x) \geq 0$ on R_+ and $h_3(x) \in L(R_+, R)$, and a number $\Lambda > 0$, such that

$$h_2{}^2(x) \leq h_1(x) h_3(x), \qquad x \in R_+, \tag{2.23}$$

$$h_1(x)\xi^2 + 2h_2(x)\xi + h_3(x) \geq \Lambda[|\tilde{\eta}(x)|^2\xi^2 + 2\,\mathrm{Re}\{\tilde{f}(x)\tilde{\eta}(-x)\}\xi$$
$$+ |\tilde{f}(x)|^2], \qquad x \in R_+, \quad \xi \in R. \tag{2.24}$$

A motivation for A_4 will follow from the discussion of a special significant case. Usually, the equations of the system (2.1) are coupled and a realistic assumption is

$$\alpha(x) = \lambda \eta(x), \qquad \lambda > 0, \qquad \int_{-\infty}^\infty \eta^2(x)\, dx > 0. \tag{2.25}$$

Then $h_1(x) = \lambda|\tilde{\eta}(x)|^2$, $h_2(x) = \lambda\,\mathrm{Re}\{\tilde{f}(x)\tilde{\eta}(-x)\}$, and taking $h_3(x) = \lambda|\tilde{f}(x)|^2$, $\Lambda = \lambda$, we see that A_2 and A_4 are verified.

Theorem 2.1   Consider the system (2.1) and assume that A_1–A_4 and H_4 hold. Then there exists at least one solution $u(t)$, $T(x, t)$ on $0 \leq t < \infty$, $x \in R$, satisfying (2.2) and such that

$$\lim_{t \to \infty} u^{(j)}(t) = 0, \qquad j = 0, 1, 2, \tag{2.26}$$

$$\lim_{t \to \infty} T(x, t) = 0, \qquad \text{uniformly in} \quad x \in R. \tag{2.27}$$

Proof   We shall start from Eq. (2.3), with $k(t)$ and $b(t)$ given by (2.14) and (2.16). After constructing $u(t)$ from (2.3), we define $T(x, t)$ by the formula (2.5).

From (2.14) and (2.16) there results

$$k^{(j)}(t) = (-1)^j \pi^{-1} \int_0^\infty x^{2j} h_1(x) \exp\{-x^2 t\} \, dx, \qquad (2.28)$$

$$b^{(j)}(t) = (-1)^j \pi^{-1} \int_0^\infty x^j h_2(x) \exp\{-x^2 t\} \, dx, \qquad (2.29)$$

for $j = 0, 1, 2, \ldots, 0 < t < \infty$. The presence of the exponential factor, A_2, and A_3 assure the convergence of both integrals for any nonnegative integer j. Moreover, (2.8) and A_2 imply $k(t) \in C(R_+, R)$ and $(-1)^j k^{(j)}(t) \geq 0, j = 0, 1, 2, \ldots, 0 < t < \infty$.

Let us now define

$$c(t) = \pi^{-1} \int_0^\infty h_3(x) \exp\{-x^2 t\} \, dx, \qquad t \in R_+. \qquad (2.30)$$

Taking into account A_4, one obtains as above

$$c^{(j)}(t) = (-1)^j \pi^{-1} \int_0^\infty x^{2j} h_3(x) \exp\{-x^2 t\} \, dx \qquad (2.31)$$

for $j = 0, 1, 2, \ldots, 0 < t < \infty$.

From (2.28)–(2.30), using the Schwarz inequality and considering A_4, we easily obtain (2.18) for $j = 0, 1, 2, \ldots$. Thus, Eq. (2.3) satisfies all the assumptions of Lemma 2.1. Therefore, (2.3) has at least one solution $u(t)$ on R_+, with $u(0) = u_0$ such that (2.26) holds. If more than one such solution exists, we choose one and denote it by $u(t)$.

Consider now the function $T(x, t)$ given by (2.5). It is constructed according to the classical results concerning the heat equation and there is no need to check the validity of the second equation of the system (2.1), nor that of the second condition (2.2). We shall prove now that the first equation of the system (2.1) is verified by the couple $u(t)$, $T(x, t)$. It is useful to notice that for each fixed t, $T(x, t)$ is square integrable on R. This property follows, for instance, from the fact that L^2 is an invariant subspace of the convolution operator with integrable kernel (see Section 2.7). From (2.5) and (2.8) we obtain

$$\tilde{T}(x, t) = f(x) \exp\{-x^2 t\} + \tilde{\eta}(x) \int_0^t \varphi(u(s)) \exp\{-x^2(t - s)\} \, ds.$$

The above formula, (2.6), (2.7), and Parseval's equality lead to

$$\int_{-\infty}^\infty \alpha(x) T(x, t) \, dx = (2\pi)^{-1} \int_{-\infty}^\infty \tilde{\alpha}(-x) \tilde{T}(x, t) \, dx$$

$$= \int_0^t k(t - s) \varphi(u(s)) \, ds + b(t) = -\dot{u}(t),$$

if we also consider (2.3). Consequently, the first equation of the system (2.1) is also verified.

The last part of the proof is concerned with (2.27) and requires further preparation. From (2.4) and (2.5) we easily obtain $T_x(x, t)$, $T_{xx}(x, t) \in L^2$, for any $t > 0$. We shall now prove the inequality

$$\sup_{x \in R} T^4(x, t) \leq 4 \int_{-\infty}^{\infty} T^2(x, t)\, dx \int_{-\infty}^{\infty} T_x{}^2(x, t)\, dx. \tag{2.32}$$

One obviously has

$$T^2(x, t) - T^2(\xi, t) = 2 \int_{\xi}^{x} T(v, t) T_x(v, t)\, dv$$

for any $x, \xi \in R$, and $t > 0$. Hence

$$|T^2(x, t) - T^2(\xi, t)|^2 \leq 4 \int_{-\infty}^{\infty} T^2(x, t)\, dx \int_{-\infty}^{\infty} T_x{}^2(x, t)\, dx. \tag{2.33}$$

Since $T(x, t) \in L^2$ for any $t > 0$, there exists a sequence $\{x_n\}$ such that

$$T(x_n, t) \to 0$$

as $n \to \infty$. This sequence depends generally on t. Now letting ξ pass through such a sequence, the inequality (2.33) leads to (2.32).

From (2.3), (2.14), and (2.16) one gets by using Fubini's theorem

$$\dot{u}(t) = -\pi^{-1} \int_0^{\infty} [h_1(x)\gamma(x, t) + h_2(x)] \exp\{-x^2 t\}\, dx, \tag{2.34}$$

with

$$\gamma(x, t) = \int_0^t \varphi(u(s)) \exp\{x^2 s\}\, ds. \tag{2.35}$$

Consider now the energy function

$$V(t) = \phi(u(t)) + (2\pi)^{-1} \int_0^{\infty} [h_1(x)\gamma^2(x, t)$$

$$+ 2h_2(x)\gamma(x, t) + h_3(x)] \exp\{-2x^2 t\}\, dx. \tag{2.36}$$

The inequality $V(t) \geq 0$ is implied by A_4 and H_4. Differentiating (2.36) and taking into account (2.34), one obtains

$$\dot{V}(t) = -\pi^{-1} \int_0^{\infty} [h_1(x)\gamma^2(x, t) + 2h_2(x)\gamma(x, t) + h_3(x)]x^2 \exp\{-2x^2 t\}\, dx,$$

$$\tag{2.37}$$

from which we see that $\dot{V}(t) \le 0$. We shall prove below that $\ddot{V}(t)$ is bounded and this property, added to $V(t) \ge 0$, $\dot{V}(t) \le 0$, will imply

$$\lim_{t \to \infty} \dot{V}(t) = 0. \tag{2.38}$$

Indeed, from $\phi(u(t)) \le V(t) \le V(0) = \phi(u_0) + (2\pi)^{-1} \int_0^\infty h_3(x)\, dx$ there results

$$|u(t)| \le \phi_0^{-1}(u_0), \tag{2.39}$$

with $\phi_0(u) \le \phi(\pm u)$, $u \in R_+$. From (2.35) and (2.39) we derive

$$|x^2 \gamma(x, t)| \le K_0, \qquad x, t \in R_+, \tag{2.40}$$

with K_0 depending only on u_0. Since

$$\ddot{V}(t) = -(2/\pi)\varphi(u(t)) \int_0^\infty [h_1(x)\gamma(x, t) + h_2(x)]x^2 \exp\{-x^2 t\}\, dx$$

$$+ (2/\pi) \int_0^\infty [h_1(x)\gamma^2(x, t) + 2h_2(x)\gamma(x, t) + h_3(x)]x^4 \exp\{-2x^2 t\}\, dx$$

and h_1, h_2, $h_3 \in L(R_+, R)$, from (2.39), (2.40), and the above formula for $\ddot{V}(t)$ we get the boundedness of the latter. Therefore (2.38) is proved.

The end of the proof of Theorem 2.1 consists in finding the estimate

$$\sup_{x \in R} T^4(x, t) \le -(8/\Lambda^2)V(0)V'(t), \tag{2.41}$$

for any $t > 0$. From (2.41) and (2.38) one obtains (2.27).

Let us remark first that (2.32) can be also written in the form

$$\sup_{x \in R} T^4(x, t) \le (1/\pi^2) \int_{-\infty}^\infty |\tilde{T}(x, t)|^2\, dx \int_{-\infty}^\infty x^2 |\tilde{T}(x, t)|^2\, dx, \tag{2.42}$$

if we apply Parseval's formula twice and recall that the Fourier transform of $T_x(x, t)$ is $-ix\tilde{T}(x, t)$. Taking into account the expression we found above for $\tilde{T}(x, t)$ and (2.35), there results

$$|\tilde{T}(x, t)|^2 = [|\tilde{\eta}(x)|^2 \gamma^2(x, t)$$

$$+ 2\,\mathrm{Re}\{\tilde{f}(x)\tilde{\eta}(-x)\}\gamma(x, t) + |\tilde{f}(x)|^2] \exp\{-2x^2 t\}\, dx,$$

which implies

$$2\pi V(0) \ge 2\pi V(t)$$

$$\ge \int_0^\infty [h_1(x)\gamma^2(x, t) + 2h_2(x)\gamma(x, t) + h_3(x)] \exp\{-2x^2 t\}\, dx$$

$$\ge \Lambda \int_0^\infty [|\tilde{\eta}(x)|^2 \gamma^2(x, t)$$

$$+ 2\,\mathrm{Re}\{\tilde{f}(x)\tilde{\eta}(-x)\}\gamma(x, t) + |\tilde{f}(x)|^2] \exp\{-2x^2 t\}\, dx$$

$$= \Lambda \int_0^\infty |\tilde{T}(x, t)|^2\, dx = (\Lambda/2) \int_{-\infty}^\infty |\tilde{T}(x, t)|^2\, dx,$$

according to A_4 and $\dot{V}(t) \le 0$. The last equality follows from the fact that $\bar{T}(-x, t) = \overline{\bar{T}(x, t)}$. We can write further

$$\int_{-\infty}^{\infty} |\tilde{T}(x, t)|^2 \, dx \le (4\pi/\Lambda)V(0), \qquad t > 0. \tag{2.43}$$

From the formula for $\dot{V}(t)$ we obtain in a similar way for $t > 0$

$$-\pi \dot{V}(t) \ge \Lambda \int_0^{\infty} x^2 |\tilde{T}(x, t)|^2 \, dx = (\Lambda/2) \int_{-\infty}^{\infty} x^2 |\tilde{T}(x, t)|^2 \, dx,$$

that is,

$$\int_{-\infty}^{\infty} x^2 |\tilde{T}(x, t)|^2 \, dx \le -(2\pi/\Lambda)\dot{V}(t), \qquad t > 0. \tag{2.44}$$

The estimate (2.41) now follows from (2.42)–(2.44). Theorem 2.1 is thereby proven.

The result established above is due to Levin and Nohel [6]. It constitutes a continuation of earlier work of these authors (see Levin and Nohel [1, 2]).

Remark 1   The system

$$\dot{u}(t) = -\int_0^{\pi} \alpha(x)T(x, t) \, dx, \qquad T_t(x, t) = T_{xx}(x, t) + \eta(x)\varphi(u(t))$$

on $0 \le x \le \pi$, $0 < t < \infty$, has been investigated by Levin and Nohel in their paper [7]. The initial conditions are $u(0) = u_0$, $T(x, 0) = f(x)$, $0 \le x \le \pi$, and the boundary conditions are $T_x(0, t) = T_x(\pi, t) = 0$, $0 < t < \infty$. It is possible to follow the same procedure because $T(x, t)$ is given by

$$T(x, t) = \int_0^{\pi} G(x, \xi; t)f(\xi) \, d\xi + \int_0^t \int_0^{\pi} G(x, \xi; t - s)\eta(\xi)\varphi(u(s)) \, d\xi \, ds$$

with

$$G(x, \xi; t) = \pi^{-1}\left(1 + 2\sum_{n=1}^{\infty} \cos nx \cos n\xi \exp\{-n^2t\}\right),$$

where $0 \le x \le \pi$, $0 < t < \infty$. Instead of using Fourier transform theory, it is convenient to handle some series occurring in the representation of various functions related to the problem.

Remark 2   The system (2.1) does not take into account the important effect of delayed neutrons.

If the special case $\varphi(u) = -1 + \exp u$ is considered and if we denote $u(t) = \log(P(t)/v)$, then (2.1) should be replaced by the system

$$\dot{P}(t) = -P(t)\int_{-\infty}^{\infty} \alpha(x)T(x, t) \, dx - (\beta/\rho)P(t) + \sum_{i=1}^{m} \lambda_i C_i(t),$$

$$\dot{C}_i(t) = (\beta_i/\rho)P(t) - \lambda_i C_i(t), \qquad i = 1, 2, \ldots, m,$$

$$T_t(x, t) = T_{xx}(x, t) + \eta(x)[P(t) - v], \qquad x \in R, \quad 0 < t < \infty.$$

The above system corresponds to an infinite continuous medium nuclear reactor with m groups of delayed neutrons having $C_i(t)$ as the concentration of the emitter in the ith group.

Fortunately, the method used in this section is applicable with minor changes. For a detailed discussion, see Levin and Nohel [8].

5.3 A Positivity Condition for the Kernel

In this section we shall deal with the integral equation

$$x(t) = h(t) + \int_0^t k(t, s)\varphi(x(s))\, ds, \qquad t \in R_+. \tag{3.1}$$

The particular case $k(t, s) = k(t - s)$ has been widely investigated in the literature, and the third chapter contains several results involving frequency conditions. As pointed out at the end of the third chapter, practically all such conditions can be viewed as positivity conditions for convenient functions or operators. We shall now establish a result concerning Eq. (3.1), the main assumption being a certain positivity property of the kernel. This result is due to Halanay [3], and in the special case of a convolution kernel it can be easily compared with Popov's type results.

Equation (3.1) can be written as $x = h + KFx$, where K is the linear Volterra integral operator generated by the kernel $k(t, s)$ and F denotes the nonlinear operator given by $(Fx)(t) = \varphi(x(t))$, $t \in R_+$. Let p, q be some positive fixed numbers and consider the linear first-order differential operator

$$B = q(d/dt) + p. \tag{3.2}$$

For any $T > 0$, let us denote

$$(x, y)_T = \int_0^T x(t)y(t)\, dt \tag{3.3}$$

for any pair of functions $x, y \in L^2([0, T], R)$. In other words, $(x, y)_T$ denotes the scalar product in the real Hilbert space $L^2([0, T], R)$.

The following definition will make precise the kind of positivity we need to obtain the existence result for Eq. (3.1). Let A be an operator from $L^2_{loc}(R_+, R)$ into itself. We shall say that A is positive and we write $A \geq 0$ if and only if for any $T > 0$ and $u \in L^2([0, T], R)$, there results $(Au, u)_T \geq 0$.

It is useful to notice that the space $L^2([0, T], R)$ can be identified with the subspace of $L^2_{loc}(R_+, R)$ consisting of all functions vanishing for $t > T$. Therefore, the above definition of positivity does not involve anything outside the space $L^2_{loc}(R_+, R)$.

If the positive operator A is such that $AL^2(R_+, R) \subset L^2(R_+, R)$, then its restriction to $L^2(R_+, R)$ is positive in the usual sense [i.e., $(Au, u) \geq 0$ for any

$u \in L^2(R_+, R)$]. This results easily from the fact that the scalar product is continuous with respect to its arguments.

Let us state now the main result of this section.

Theorem 3.1 Consider Eq. (3.1) under the following assumptions.

1. $k(t, s)$ is continuous from $D = \{(t, s) : 0 \leq s \leq t < \infty\}$ into R and $k_t(t, s)$ is locally square integrable in D;

2. $\varphi(x)$ is a continuous mapping from R into itself such that

$$-\lambda_1 x^2 \leq x\varphi(x) \leq -\lambda_0 x^2 \qquad (3.4)$$

for any $x \in R$, with $\lambda_0 > 0$;

3. $h(t)$, $\dot{h}(t) \in L^2(R_+, R)$;

4. there exist positive numbers p, q, and $\lambda > \lambda_1$ such that the positivity condition

$$BK + (p/\lambda)I \geq 0 \qquad (3.5)$$

holds true, where I denotes the identity operator.

Then, any continuous solution on R_+ of (3.1) is such that

$$x(t) \in C(R_+, R) \cap L^2(R_+, R). \qquad (3.6)$$

Proof Let us remark first that (3.5) has a meaning only if we prove that BK acts from $L^2_{loc}(R_+, R)$ into itself. Since $(BKu)(t) = q(t, t)u(t) + q \int_0^t k_t(t, s)u(s) \, ds + p \int_0^t k(t, s)u(s) \, ds$, the needed property follows readily from condition (1) and the Schwarz inequality for integrals.

As remarked above, Eq. (3.1) can be written as $x = h + KFx$. From this, we get $Bx = Bh + BKFx \in L^2_{loc}(R_+, R)$. For any $T > 0$, we have

$$(Bx, Fx)_T = (BKFx, Fx)_T + (Bh, Fx)_T. \qquad (3.7)$$

But

$$(Bx, Fx)_T = q \int_0^T \varphi(x(t))\dot{x}(t) \, dt + p \int_0^T \varphi(x(t))x(t) \, dt$$

$$= q \int_{x(0)}^{x(T)} \varphi(\xi) \, d\xi + p \int_0^T \varphi(x(t))x(t) \, dt$$

$$= q \int_0^{x(T)} \varphi(\xi) \, d\xi + p \int_0^T \varphi(x, t))x(t) \, dt - q \int_0^{x(0)} \varphi(\xi) \, d\xi.$$

Condition (2) implies

$$-(1/\lambda_0)\varphi^2(x) \leq x\varphi(x) \leq -(1/\lambda_1)\varphi^2(x)$$

and

$$-(\lambda_1/2)x^2 \leq \int_0^x \varphi(\xi) \, d\xi \leq -(\lambda_0/2)x^2.$$

Consequently,

$$(Bx, Fx)_T \leq -(p/\lambda_1) \int_0^T \varphi^2(x(t)) \, dt + (q\lambda_1/2)x^2(0) - (q\lambda_0/2)x^2(T).$$

Therefore, (3.7) yields

$$(BKFx, Fx)_T + (Bh, Fx)_T \leq -(p/\lambda_1)(Fx, Fx)_T + (q\lambda_1/2)x^2(0) - (q\lambda_0/2)x^2(T).$$

According to our assumption (4), one can write

$$(BKFx, Fx)_T + (p/\lambda)(Fx, Fx) \geq 0.$$

Comparing the last two inequalities one obtains

$$(p(\lambda - \lambda_1)/\lambda\lambda_1)(Fx, Fx)_T \leq (q\lambda_1/2)x^2(0) + |(Bh, Fx)_T| - (q\lambda_0/2)x^2(T),$$

which leads to

$$\alpha \|Fx\|_T^2 \leq \beta + \|Bh\|_T \|Fx\|_T, \tag{3.8}$$

where α and β are positive numbers. Condition (3) shows that $\|Bh\|_T \leq M < \infty$ for any $T > 0$. Hence, (3.8) implies $\|Fx\|_T \leq M_1 < \infty$ for any $T > 0$. This means that $Fx \in L^2(R_+, R)$. But (3.4) implies $|x(t)| \leq \lambda_0^{-1}|(Fx)(t)|$, and this shows that $x \in L^2(R_+, R)$. Finally, the inequality from which we derived (3.8) yields

$$(q\lambda_0/2)x^2(T) \leq (q\lambda_1/2)x^2(0) + \|Bh\|_T \|Fx\|_T - (p(\lambda - \lambda_1)/\lambda\lambda_1)\|Fx\|_T^2.$$

This inequality together with $Fx \in L^2(R_+, R)$ leads to the conclusion that $|x(t)|$ is bounded on R_+. Theorem 3.1 is thus proven.

Remark 1   We started with a continuous solution of (3.1). If we assume that there exists a solution $x(t)$ of (3.1) which belongs to $L^\infty_{\text{loc}}(R_+, R)$, then it follows easily $x(t) \in C_c(R_+, R)$.

Remark 2   The assumption on $\varphi(x)$ constitutes a sector-type restriction. It could be replaced by the usual one $\lambda_0 x^2 \leq x\varphi(x) \leq \lambda_1 x^2$, $\lambda_0 > 0$, provided we change (3.5) to a condition of negativity (the equation remains unchanged if we put $-k(t, s)$ and $-\varphi(x)$ instead of $k(t, s)$ and $\varphi(x)$, respectively).

Remark 3   If we allow $q = 0$ in the statement, then only $x \in L^2(R_+, R)$ can be derived.

We shall now consider the equation

$$x(t) = h(t) + \int_0^t k(t - s)\varphi(x(s)) \, ds, \tag{3.9}$$

with the objective of finding a frequency condition that substitutes for the positivity condition (3.5).

Let us remark first that the positivity condition (3.5) becomes in our case

$$\int_0^T u(t)\left\{\int_0^t [qk(t-s) + pk(t-s)]u(s)\,ds\right\} dt$$

$$+ [qk(0) + (p/\lambda)]\int_0^T u^2(t)\,dt \geq 0, \qquad (3.10)$$

for any $u \in L^2([0, T], R)$ and $T > 0$.

Second, if instead of condition (1) we assume that $k(t)$, $\dot{k}(t) \in L(R_+, R)$, there results easily the fact that BK acts from $L^2_{loc}(R_+, R)$ into itself. Moreover, BK carries $L^2(R_+, R)$ into itself, and this feature suggests to us that the positivity condition should have a simpler form.

We are going to prove that (3.10) is equivalent to the Popov's type condition

$$\text{Re}\{(p - isq)\tilde{k}(s)\} + (p/\lambda) \geq 0 \qquad (3.11)$$

for any $s \in R$. Indeed, if we denote by $u_T(t)$ the truncation of $u(t)$ and by $v_T(t)$ the convolution product of the kernel Bk with u_T, then

$$v_T(t) = \int_0^t [q\dot{k}(t-s) + pk(t-s)]u_T(s)\,ds$$

for $t \in R_+$. It is obvious that $v_T \in L(R_+, R) \cap L^2(R_+, R)$ and we have $\tilde{v}_T(s) = [(p - isq)\tilde{k}(s) - qk(0)]\tilde{u}_T(s)$. Using Parseval's formula we obtain

$$\int_0^T u(t)v_T(t)\,dt = \int_0^\infty u_T(t)v_T(t)\,dt$$

$$= (2\pi)^{-1}\int_{-\infty}^\infty \text{Re}\{\overline{\tilde{u}_T(s)}\tilde{v}_T(s)\}\,ds$$

$$= (2\pi)^{-1}\int_{-\infty}^\infty \text{Re}\{(p - isq)\tilde{k}(s) - qk(0)\}|\tilde{u}_T(s)|^2\,ds.$$

On the other hand, going back to (3.10) we notice that

$$[qk(0) + p/\lambda]\int_0^T u^2(t)\,dt = [qk(0) + (p/\lambda)]\int_0^\infty u_T^2(t)\,dt$$

$$= (2\pi)^{-1}[qk(0) + (p/\lambda)]\int_{-\infty}^\infty |\tilde{u}_T(s)|^2\,ds.$$

Therefore

$$\int_0^T u(t)v_T(t)\,dt + [qk(0) + (p/\lambda)]\int_0^T u^2(t)\,dt$$

$$= (2\pi)^{-1}\int_{-\infty}^\infty \text{Re}\{(p - isq)\tilde{k}(s) + (p/\lambda)\}|\tilde{u}_T(s)|^2\,ds,$$

which can be also written as

$$\int_0^T u(t) \left\{ \int_0^t [q\dot{k}(t-s) + pk(t-s)]u(s)\, ds \right\} dt + [qk(0) + (p/\lambda)] \int_0^T u^2(t)\, dt$$

$$= (2\pi)^{-1} \int_{-\infty}^{\infty} \text{Re}\{(p - isq)\tilde{k}(s) + (p/\lambda)\} |u_T(s)|^2\, ds.$$

If we compare this equality with (3.10), we see that (3.10) and (3.11) are equivalent. We arrive at this conclusion by considering that any function which is continuously differentiable of the second order and vanishes outside a compact interval is the Fourier transform of a certain function belonging to L (the inversion formula holds true).

Actually, our assumptions concerning Eq. (3.9) imply $\lim x(t) = 0$ as $t \to \infty$. Indeed, we obtain from (3.9) by differentiation

$$\dot{x}(t) = \dot{h}(t) + k(0)\varphi(x(t)) + \int_0^t \dot{k}(t-s)\varphi(x(s))\, ds.$$

As seen in the proof of Theorem 3.1, $\varphi(x(t)) \in L^2$. Therefore, $\dot{x}(t) \in L^2$. Since $x(t)$ is bounded on R_+, there results that $f(t) = x^2(t)$ is uniformly continuous and integrable on R_+. Barbălat's lemma gives the needed result. In other words, (3.6) can be replaced by the more accurate relation

$$x(t) \in C_0(R_+, R) \cap L^2(R_+, R).$$

In concluding this section we notice that under appropriate positivity assumptions concerning the kernel, it is possible to obtain other kinds of behavior for the solution. In particular, as pointed out by Halanay [3], the positivity condition used by Barbu (see Section 3.8) can be easily generalized to nonconvolution kernels.

5.4 Linearization of Volterra Integral Equations

Consider the system of integral equations

$$x(t) = h(t) + \int_0^t k(t-s)\varphi(x(s))\, ds, \qquad t \in R_+, \qquad (4.1)$$

where x, h, and φ are real n-vectors, $k(t)$ is an n by n matrix kernel, and $\varphi(0) = 0$.

If we assume that $\varphi(x)$ is continuously differentiable of the first order at $x = 0$, then the jacobian matrix $J = (\partial\varphi_i/\partial x_j|_{x=0})$ exists and the linearized system

$$y(t) = h(t) + \int_0^t k(t-s)Jy(s)\, ds, \qquad t \in R_+, \qquad (4.2)$$

can be associated with (4.1). Since $\varphi(x) - Jx = O(\|x\|)$ for small $\|x\|$, we can expect that for small $h(t)$ the behavior of solutions would be the same for both Eq. (4.1) and (4.2).

An answer to the linearization problem has been given recently by Miller [5]. In order to prove this result, we need a lemma that is interesting in itself.

Lemma 4.1 Consider the system (4.1) and assume that the following conditions hold:

1. $h(t) \in C(R_+, R^n)$ and is uniformly continuous on R_+;
2. $\|k(t)\| \in L(R_+, R)$;
3. $\varphi(x)$ is a continuous mapping from R^n into itself.

Then, if $x(t)$ is a solution of (4.1) belonging to $C(R_+, R^n)$ and Ω is its positive limit set, the following property holds: to each point $z \in \Omega$, there corresponds a sequence $\{t_m\}$, with $t_m \to \infty$ as $m \to \infty$, and functions $X(t)$, $H(t)$ such that $X(t) \in \Omega$ for $t \in R$,

$$\lim_{m \to \infty} (\|x(t + t_m) - X(t)\| + \|h(t + t_m) - H(t)\|) = 0$$

uniformly on any compact subset of R, and

$$X(t) = H(t) + \int_{-\infty}^{t} k(t - s)\varphi(X(s))\, ds, \qquad t \in R. \tag{4.3}$$

Proof Let us remark first that our assumptions imply the fact that $x(t)$ is uniformly continuous on R_+. It suffices to observe that the convolution product of a function from $C(R_+, R)$ by a function from $L(R_+, R)$ is uniformly continuous.

If we fix $z \in \Omega$, then there exists a sequence $\{t_m\}$, with $t_m \to \infty$, $x(t_m) \to z$ as $m \to \infty$. Define now $x_m(t) = x(t + t_m)$, $h_m(t) = h(t + t_m)$ for $t \geq -t_m$, $m = 1$, $2, \ldots$. It follows easily from (4.1) that

$$x_m(t) = h_m(t) + \int_{-t_m}^{t} k(t - s)\varphi(x_m(s))\, ds, \tag{4.4}$$

$t \geq -t_m$, $m = 1, 2, \ldots$. According to our assumption $x(t) \in C(R_+, R^n)$, the sequence $\{x_m(t)\}$ is uniformly bounded. From (4.4) and conditions (1)–(3) we get without any difficulty that $\{x_m(t)\}$ is also equicontinuous (on any half-axis $t \geq -T$). Therefore, we can extract a subsequence of $\{x_m(t)\}$ that is uniformly convergent on any compact subset of R to a continuous function $X(t)$. There is no loss of generality if we assume that $\{x_m(t)\}$ itself converges to $X(t)$. At this point, (4.4) can be used in order to show that $\{h_m(t)\}$ also converges to a certain continuous function $H(t)$ and then to get (4.3).

It remains to show that $X(t) \in \Omega$, $t \in R$. Indeed, for fixed t, $X(t) = \lim x_m(t) = \lim x(t + t_m)$ as $m \to \infty$. If we denote $t + t_m = t_m'$, we have $t_m' \to \infty$ as $m \to \infty$ and the lemma is proven.

Let us mention that Lemma 4.1 will not be directly applied to the system (4.1). First, we shall write (4.1) in an equivalent form using the resolvent kernel.

Theorem 4.1 Consider the system (4.1) and assume that the following conditions hold:

1. $h(t) \in C(R_+, R^n)$;
2. $k(t)$ is locally integrable on R_+;
3. $\varphi(x)$ is continuously differentiable of the first order from R^n into itself and $\varphi(0) = 0$;
4. the jacobian matrix $\varphi'(x) = (\partial \varphi_i / \partial x_j)$ does not vanish at $x = 0$ (i.e., $J \neq 0$);
5. the resolvent kernel associated with $k(t)J$, say $r(t)$, exists on R_+ and is such that $\|r(t)\| \in L(R_+, R)$.

Then there exists $\varepsilon_0 > 0$ with the property: if $\varepsilon < \varepsilon_0$ and $\|y(t)\| \le \varepsilon/2$ on R_+, where $y(t)$ denotes the solution of (4.2), we have $\|x(t)\| \le \varepsilon$ on R_+. Moreover, if $\|y(t)\| \to 0$ as $t \to \infty$, then $\|x(t)\| \to 0$ as $t \to \infty$.

Proof Let us recall first that $r(t)$ satisfies the equation

$$r(t) = k(t)J + \int_0^t k(t - s)Jr(s)\,ds, \qquad t \in R_+. \tag{4.5}$$

From (4.1), (4.2), and (4.5), we obtain

$$x(t) = y(t) + \int_0^t r(t - s)\psi(x(s))\,ds, \qquad t \in R_+, \tag{4.6}$$

where

$$\psi(x) = \varphi(x) - Jx = O(\|x\|), \qquad \|x\| \to 0. \tag{4.7}$$

It is easy to check that (4.6) is equivalent to (4.1).

Taking into account (4.7), we can find $\varepsilon_0 > 0$ such that for any $\varepsilon < \varepsilon_0$, from $\|x\| \le \varepsilon$ there results

$$2\|\psi(x)\| \int_0^\infty \|r(s)\|\,ds \le \|x\|, \qquad 2\|\varphi'(x) - J\| \int_0^\infty \|r(s)\|\,ds \le 1.$$

In the space $C(R_+, R^n)$ we consider the ball $S_\varepsilon = \{x \mid x \in C, |x|_C \le \varepsilon\}$. The operator T given by

$$(Tx)(t) = y(t) + \int_0^t r(t - s)\psi(x(s))\,ds, \qquad t \in R_+,$$

carries S_ε into itself and is a contraction mapping. The first part of the theorem is thereby proven.

Let us now prove that $\|y(t)\| \to 0$ as $t \to \infty$ implies the same property for $x(t)$. At this point Lemma 4.1 provides the necessary tool.

Let Ω be the positive limit set of the solution $x(t)$. Since $x(t)$ is bounded, it follows that Ω is nonempty, compact, and connected. Using Lemma 4.1 and taking into account the assumption $\|y(t)\| \to 0$ as $t \to \infty$, there results that Ω is the union of all solutions of

$$z(t) = \int_{-\infty}^{t} k(t-s)\psi(z(s))\,ds, \qquad t \in R, \tag{4.8}$$

with $\|z(t)\| \le \varepsilon$, $t \in R$. But in the space $C(R, R)$, Eq. (4.8) has the only solution $z(t) \equiv 0$ (more precisely, the uniqueness holds in the ball of radius ε, centered at the origin). This readily follows by the contraction mapping principle. Therefore, Ω reduces to the unique point that is the origin in R^n. Hence $\|x(t)\| \to 0$ as $t \to \infty$ and Theorem 4.1 is proven.

Remark 1   If we assume $\|k(t)\| \in L(R_+, R)$, then condition (5) in the statement of Theorem 4.1 takes the equivalent form

$$\det(I - \tilde{k}(s)) \ne 0, \qquad \text{Im } s \ge 0. \tag{4.9}$$

This statement follows from Banach algebra arguments we are not going to develop here.

Remark 2   The second half of the theorem can be proved by using the contraction mapping principle for Eq. (4.6) in the space $C_0(R_+, R^n)$. One has to consider that $\varphi(0) = 0$ and that the convolution operator with integrable matrix kernel maps $C_0(R_+, R^n)$ into itself (see Section 2.7). This last approach shows that the linearization problem is a special case of the perturbation problem.

Nevertheless, the proof given above deserves our attention due to the fact that Lemma 4.1 is involved. Such concepts as limit set or limit equation are of current interest in the investigation of the behavior of solutions.

5.5   Volterra–Stieltjes Integral Equations

We shall now consider the equation

$$\dot{x}(t) = h(t) - \int_0^t \varphi(x(t-s))\,dB(s), \qquad t \in R_+, \tag{5.1}$$

which constitutes an immediate generalization of the equation

$$\dot{x}(t) = h(t) - \int_0^t \varphi(x(t-s))b(s)\,ds, \qquad t \in R_+. \tag{5.2}$$

Indeed, it suffices to put $B(t) = \int_0^t b(s)\, ds$ in order to write (5.2) in the form (5.1).

The assumptions under which the results of this section will be established vary from theorem to theorem. A hypothesis we shall keep throughout this section is the following one:

H_B: $B(t)$ is of bounded variation on any interval $[0, T]$, $T > 0$, $B(0) = 0$, and $B(t) = \frac{1}{2}[B(t + 0) + B(t - 0)]$.

The main attention will be paid to the boundedness of solutions of (5.1). Besides boundedness we shall also consider some oscillation properties.

Since no result concerning the existence of solutions for (5.1) was proved in the preceding chapters, we shall deal first with a general existence theorem.

Theorem 5.1   Consider Eq. (5.1) and assume that the following conditions are fulfilled:

1. $h(t) \in L_{\mathrm{loc}}(R_+, R)$;
2. $\varphi(x)$ is a locally Lipschitzian map from R into itself such that

$$\varphi(x) \geq -\lambda, \qquad x \in R, \tag{5.3}$$

with $\lambda > 0$ and

$$\sup \varphi(x) < +\infty, \qquad -\infty < x \leq 0; \tag{5.4}$$

3. $B(t)$ satisfies H_B and $B(t) \geq 0$.

Then there exists a unique solution $x(t)$ of (5.1) that is absolutely continuous on any compact subset of R_+ such that $x(0) = x_0 \in R$.

Proof   The integral version of Eq. (5.1) has the form

$$x(t) = x_0 + H(t) - \int_0^t \varphi(x(s))B(t - s)\, ds, \tag{5.5}$$

where $H(t) = \int_0^t h(s)\, ds$, $t \in R_+$. Indeed, if we assume that $x(t)$, with $x(0) = x_0$, satisfies (5.1) and is absolutely continuous on any compact subset of R_+, the integration yields

$$x(t) = x_0 + H(t) - \int_0^t \left\{ \int_0^s \varphi(x(s - u))\, dB(u) \right\} ds. \tag{5.6}$$

Using Fubini's theorem and the formula of integration by parts for the Riemann–Stieltjes integral, we obtain

$$\int_0^t \left\{ \int_0^s \varphi(x(s-u))\, dB(u) \right\} ds = \int_0^t \left\{ \int_u^t \varphi(x(s-u))\, ds \right\} dB(u)$$

$$= \int_0^t \left\{ \int_0^{t-u} \varphi(x(v))\, dv \right\} dB(u)$$

$$= \int_0^t B(u)\, d\left\{ \int_0^{t-u} \varphi(x(v))\, dv \right\}$$

$$= -\int_0^t \varphi(x(t-u))B(u)\, du.$$

This shows that (5.6) is equivalent to (5.5). Conversely, if we assume that $x(t)$ is a continuous solution for (5.5), we get easily that it satisfies (5.1) and $x(0) = x_0$.

Define now the successive approximations for (5.5) by

$$x_0(t) = x_0 + H(t), \qquad t \in R_+,$$

$$x_{n+1}(t) = x_0(t) - \int_0^t \varphi(x_n(s))B(t-s)\, ds, \qquad t \in R_+, \quad n = 1, 2, \ldots. \tag{5.7}$$

We shall now prove that the sequence $\{x_n(t)\}$ is uniformly convergent on any compact interval $[0, T]$ to a (continuous) function $x(t)$ that satisfies (5.5).

Let $T > 0$ be given. According to our assumptions, $|H(t)| \le H_0 = H_0(T)$ on $[0, T]$. Therefore

$$x_{n+1}(t) \le x_0 + H_0 + \lambda \int_0^T B(s)\, ds = C_0, \qquad n = 1, 2, \ldots$$

on $[0, T]$. This implies

$$|\varphi(x_n(t))| \le C_1, \qquad n = 0, 1, 2, \ldots$$

on $[0, T]$, and the last inequality leads to

$$|x_{n+1}(t) - x_0| \le H_0 + C_1 \int_0^T B(s)\, ds = C_2, \qquad n = 1, 2, \ldots.$$

In other words, the sequence $\{x_n(t)\}$ is uniformly bounded on $[0, T]$. Namely, we can write $|x_n(t)| \le |x_0| + C_2 = C_3$ on $[0, T]$ for any $n \ge 1$. If we denote by $L = L(C_3)$ the Lipschitz constant corresponding to $\varphi(x)$ on $|x| \le C_3$, and by B_0 the upper bound of $B(t)$ on $[0, T]$, then (5.7) leads to

$$|x_{n+1}(t) - x_n(t)| \le B_0 L \int_0^t |x_n(s) - x_{n-1}(s)|\, ds, \qquad t \in [0, T], \quad n = 1, 2, \ldots.$$

$$\tag{5.8}$$

It is now obvious that $\{x_n(t)\}$ converges uniformly on $[0, T]$ to a continuous solution $x(t)$ of (5.5).

The uniqueness follows by the same standard way of successive approximations or by using Gronwall's lemma.

Corollary Let us keep assumptions (1) and (3) of Theorem 5.1 and replace assumption (2) by the following one: $\varphi(x)$ is a continuous map from R into itself such that (5.3) and (5.4) hold true. Then there exists at least one solution $x(t)$ of (5.1), absolutely continuous on any interval $[0, T]$, $T > 0$, such that $x(0) = x_0$.

Indeed, we can construct a sequence $\{\varphi_n(x)\}$ of locally Lipschitzian maps from R into itself such that $\{\varphi_n(x)\}$ converges uniformly to $\varphi(x)$ on any compact subset of R and each $\varphi_n(x)$ satisfies both (5.3) and (5.4), the last one with the same upper bound for all n. The equation

$$\dot{x}(t) = h(t) - \int_0^t \varphi_n(x(t - s))\, dB(s), \qquad t \in R_+, \tag{5.9}$$

has a unique solution $x_n(t)$ such that $x_n(0) = x_0$. Using also the integral version of (5.9), it can be easily seen that $\{x_n(t)\}$ is uniformly bounded and equicontinuous on any interval $[0, T]$, $T > 0$. Therefore, there exists a subsequence $\{x_{n_k}(t)\}$, uniformly convergent on any $[0, T]$, $T > 0$. The limit of such a subsequence is obviously a solution of (5.1) under conditions stated in the corollary.

Before stating the first boundedness result for Eq. (5.1), we prove the following lemma concerning measurable functions.

Lemma 5.1 Let $f(t)$ be a measurable real-valued function on R_+. The following two properties are considered:

    i. for each $\varepsilon > 0$, mes$\{t : |f(t)| > \varepsilon\} < \infty$;

    ii. for each $\varepsilon > 0$, there exists $T(\varepsilon) > 0$ such that $\Omega \subset [T, \infty)$ and mes $\Omega \geq m_0$ imply inf $|f(t)| < \varepsilon$ on Ω, with $m_0 > 0$ a prescribed constant.

If $f(t)$ satisfies (i), then it satisfies (ii) for every $m_0 > 0$. If $f(t)$ satisfies (ii) for some $m_0 > 0$, then $f(t)$ satisfies (i).

Proof Assume that (i) holds and let $\varepsilon > 0$, $m_0 > 0$ be given. Then mes$\{t : |f(t)| > \varepsilon/2\} < \infty$. This shows that for some $T = T(\varepsilon)$,

$$\text{mes}(\{t : |f(t)| > \varepsilon/2\} \cap [T, \infty)) < m_0.$$

For any measurable set $\Omega \subset [T, \infty)$, with mes $\Omega \geq m_0$, there exists $t \in \Omega$ such that $|f(t)| \leq \varepsilon/2$. Therefore, inf $|f(t)| < \varepsilon$ on Ω.

Assume now that (ii) holds for a certain $m_0 > 0$ and

$$\text{mes}\{t : |f(t)| > \varepsilon_1\} = \infty.$$

for some $\varepsilon_1 > 0$. Let Ω_1 be the set $\{t : |f(t)| > \varepsilon_1\} \cap [T_1, \infty)$, where $T_1 = T(\varepsilon_1)$. Then $\Omega_1 \subset [T_1, \infty)$ and mes $\Omega_1 = \infty > m_0$. Consequently, (ii) implies inf $|f(t)| < \varepsilon_1$ on Ω_1. On the other hand, from the definition of Ω_1 we get inf $|f(t)| \geq \varepsilon_1$ on Ω_1. The contradiction proves the lemma.

Theorem 5.2 Assume that the following conditions hold for (5.1):

1. $h(t) \in L^\infty(R_+, R)$ and for each $\varepsilon > 0$, mes$\{t : |h(t)| > \varepsilon\} < \infty$;
2. $\varphi(x)$ is a continuous map from R into itself, $\varphi(x) \geq -\lambda$ for $x \in R$ with given $\lambda > 0$, and

$$\liminf_{x \to \infty} \varphi(x) > 0, \qquad \limsup_{x \to -\infty} \varphi(x) < 0;$$

3. $B(0) = 0$, $B(t)$ is nondecreasing on R_+, and $0 < B(\infty) < \infty$.

If $x(t)$ is a solution of (5.1) on R_+, then it is bounded.

Remark The conditions of Theorem 5.2 are obviously stronger than those of the corollary to Theorem 5.1. Therefore, for any $x_0 \in R$ there exists at least one solution of (5.1) on R_+ such that $x(0) = x_0$.

Proof of Theorem 5.2 It follows readily from (5.1) and our hypotheses that any solution $x(t)$ given on R_+ satisfies

$$\dot{x}(t) \leq \lambda B(\infty) + |h|_{L^\infty} = c_1 < \infty \qquad \text{a.e. on } R_+. \tag{5.10}$$

Let us now prove that $x(t)$ is bounded from above on R_+, i.e.,

$$\limsup_{t \to \infty} x(t) < \infty. \tag{5.11}$$

Assume that (5.11) does not hold. Let $y_n = x(0) + nc_1$, $n = 0, 1, 2, \ldots$. From our assumptions there results the existence of a sequence $\{s_n\}$ that is uniquely determined by $s_0 = 0$, $x(t) < x(s_n) = y_n$, $0 \leq t < s_n$, $n = 1, 2, \ldots$. The sequence $\{\xi_n\}$ is also uniquely defined by

$$y_{n-1} = x(\xi_n) < x(t) < x(s_n) = y_n, \qquad s_{n-1} \leq \xi_n < t < s_n. \tag{5.12}$$

Denote by I the subset of R_+ consisting of all t such that $\dot{x}(t)$ exists and

$$0 \leq \dot{x}(t) \leq \lambda B(\infty) + |h|_{L^\infty}.$$

Taking into account (5.10) and the preceding definitions we get

$$c_1 = y_n - y_{n-1} = \int_{\xi_n}^{s_n} \dot{x}(t) \, dt \leq c_1 \, \mathrm{mes}\{[\xi_n, s_n] \cap I\},$$

from which

$$\mathrm{mes}\{[\xi_n, s_n] \cap I\} \geq 1, \qquad n = 1, 2, \ldots. \tag{5.13}$$

From (5.13), Lemma 5.1, and condition (1) of the theorem, we derive the existence of a sequence $\{t_n\}$ such that

$$t_n \in [\xi_n, s_n] \cap I, \qquad \lim_{n \to \infty} h(t_n) = 0. \tag{5.14}$$

Since $y_n \to \infty$ as $n \to \infty$, from (5.12) there results $x(t_n) \to \infty$ as $n \to \infty$. Define now

$$T_n = \min\{t_n, (1/2c_1)x(t_n)\}, \qquad n = 1, 2, \ldots.$$

Then $T_n \to \infty$ as $n \to \infty$ and

$$x(t_n) - x(t) = \int_t^{t_n} \dot{x}(s)\, ds \le c_1(t_n - t) \le \tfrac{1}{2}x(t_n)$$

for $t_n - T_n \le t \le t_n$. In other words,

$$x(t) \ge \tfrac{1}{2}x(t_n), \qquad t_n - T_n \le t \le t_n, \quad n = 1, 2, \ldots.$$

Let us denote $\lambda_1 = \liminf \varphi(x)$ as $x \to \infty$, if $\liminf \varphi(x) < \infty$ as $x \to \infty$, and $\lambda_1 = 1$ otherwise. Then there exists an integer N such that

$$\varphi(x(t)) \ge \tfrac{1}{2}\lambda_1 > 0, \qquad t_n - T_n \le t \le t_n, \quad n \ge N. \tag{5.15}$$

From (5.10), (5.14), (5.15), $\varphi(x) \ge -\lambda$, and the definition of I, one gets for $n \ge N$

$$\dot{x}(t_n) = -\int_0^{T_n} \varphi(x(t_n - s))\, dB(s) - \int_{T_n}^{t_n} \varphi(x(t_n - s))\, dB(s)$$

$$\dot{x}(t_n) = -\int_0^{T_n} \varphi(x(t_n - s))\, dB(s) - \int_{T_n}^{t_n} \varphi(x(t_n - s))\, dB(s) + h(t_n)$$

$$\le -\tfrac{1}{2}\lambda_1 B(T_n) + \lambda[B(\infty) - B(T_n)] + h(t_n).$$

Since $B(\infty) > 0$ and $T_n \to \infty$ as $n \to \infty$, the last inequality and (5.14) yield $\dot{x}(t_n) < 0$ for sufficiently large n. But $t_n \in I$ implies $\dot{x}(t_n) \ge 0$ for all n. The contradiction proves (5.11).

Condition (2) of the theorem and (5.11) show that

$$\sup_{t \in R_+} |\varphi(x(t))| < \infty. \tag{5.16}$$

We now define the functions $\psi(y)$ and $y(t)$ as follows:

$$\psi(y) = -\varphi(-y), \qquad y \in R, \qquad y(t) = -x(t), \qquad t \in R_+. \tag{5.17}$$

Then

$$\liminf_{y \to \infty} \psi(y) > 0 \tag{5.18}$$

according to the hypothesis, and (5.16) yields lim sup $|\psi(y(t))| < \infty$ as $t \to \infty$. For some positive constant c_2 we can write

$$\psi(y(t)) \geq -c_2 > -\infty, \qquad t \in R_+. \tag{5.19}$$

From (5.17) and (5.1) there results

$$\dot{y}(t) = -h(t) - \int_0^t \psi(y(t-s))\, dB(s), \tag{5.20}$$

which shows that $y(t)$ satisfies an equation of the form (5.1). Taking into account our assumptions, one sees that the same arguments used in establishing (5.11) lead to

$$\limsup_{t \to \infty} y(t) < \infty. \tag{5.21}$$

From (5.17) and (5.21) we get

$$\liminf_{t \to \infty} x(t) > -\infty, \tag{5.22}$$

which together with (5.11) end the proof of Theorem 5.2.

Example   Let us consider Eq. (5.1) with

$$h(t) = (1+t)^{-1} + t\exp\{-t\}, \qquad t \in R_+,$$

$$\varphi(x) = \begin{cases} -1, & x < -1, \\ x, & -1 \leq x \leq 1, \\ \exp\{1 - \exp[x-1]\}, & x \geq 1, \end{cases}$$

$B(t) = \int_0^t b(s)\, ds$, where $b(t) = \exp\{-t\}$, $t \in R_+$. By direct substitution one sees that $x(t) = 1 + \ln(1+t)$ is an unbounded solution. This occurs because the condition $\varphi(x) \geq -\lambda$ is violated.

The next result concerns the oscillation property of unbounded solutions of Eq. (5.1), under suitable conditions.

Theorem 5.3   Consider Eq. (5.1) and assume that conditions (1) and (3) of Theorem 5.2 are verified. Let $\varphi(x)$ be a continuous map from R into itself such that lim inf $\varphi(x) > 0$ as $x \to \infty$ and lim sup $\varphi(x) < 0$ as $x \to -\infty$. If $x(t)$ is a solution on R_+ of (5.1) and $\sup |x(t)| = \infty$ on R_+, then

$$\limsup_{t \to \infty} x(t) = -\liminf_{t \to \infty} x(t) = \infty. \tag{5.23}$$

Proof   Suppose that lim sup $x(t) = \infty$ as $t \to \infty$ and

$$\liminf_{t \to \infty} x(t) > -\infty. \tag{5.24}$$

Our assumptions on $\varphi(x)$ and (5.24) imply

$$\varphi(x(t)) \geq -\lambda > -\infty, \qquad t \in R_+, \tag{5.25}$$

for suitable $\lambda > 0$. From (5.25) and conditions (1) and (3) of Theorem 5.2, we derived lim sup $x(t) < +\infty$ as $t \to \infty$. In other words, (5.24) leads to a contradiction. Hence, (5.23) is verified.

If lim inf $x(t) = -\infty$ as $t \to \infty$ but

$$\limsup_{t \to \infty} x(t) < \infty, \tag{5.26}$$

then we define $\psi(y) = -\varphi(-y)$, $y(t) = -x(t)$, and proceed as above in order to show that (5.26) cannot hold. This completes the proof.

The next result is concerned with boundedness of solutions of Eq. (5.1), dropping the assumption that $B(t)$ is nondecreasing. More precisely, the following statement holds.

Theorem 5.4 Let us consider Eq. (5.1) under the following conditions:

1. $h(t) \in L_{\text{loc}}(R_+, R)$ and $\sup_{t \in R_+} |\int_0^t h(s)\, ds| < \infty$;
2. $\varphi(x)$ is a continuous map from R into itself, $\varphi(x) \geq -\lambda$ on R_+ with $\lambda > 0$, and $x\varphi(x) > 0$ for $|x| \geq X$, $0 \leq X < \infty$, X being fixed;
3. $B(0) = 0$, $B(t) \geq 0$ and has bounded variation on R_+, $B(\infty) > 0$, and $B(t)$ is either nondecreasing or nonincreasing on $[T, \infty)$, $T \geq 0$; in the last case the inequality $B(t) > B(\infty)$ is verified on $[T, \infty)$;
4. the moment condition

$$\left| \int_0^\infty t\, dB(t) \right| < \infty \tag{5.27}$$

is satisfied.

Then any solution $x(t)$ of (5.1) existing on R_+ is bounded there.

Proof Let us remark first that conditions (3) and (4) imply

$$M = \int_0^\infty |B(\infty) - B(t)|\, dt < \infty. \tag{5.28}$$

It suffices to show that

$$\int_T^\infty [B(\infty) - B(t)]\, dt < \infty, \tag{5.29}$$

if we assume $B(t)$ nondecreasing on $[T, \infty)$. From (5.27) there results (for $\tau \geq T$)

$$\int_\tau^\infty t\, dB(t) \geq \tau \int_\tau^\infty dB(t) = \tau[B(\infty) - B(\tau)] \to 0$$

as $\tau \to \infty$. On the other hand

$$\int_T^\tau [B(\infty) - B(t)]\, dt = \tau[B(\infty) - B(\tau)] - T[B(\infty) - B(T)] + \int_T^\tau t\, dB(t).$$

For $\tau \to \infty$ we obtain

$$\int_T^\infty [B(\infty) - B(t)]\, dt = -T[B(\infty) - B(T)] + \int_T^\infty t\, dB(t),$$

which proves (5.29). A similar argument holds when $B(t)$ is nonincreasing.

We shall now consider the case when $B(t)$ is nondecreasing on $[T, \infty)$. Define

$$H(t) = \int_0^t h(s)\, ds, \qquad H_0 = \sup_{t \in R_+} |H(t)|, \qquad B_0 = \sup_{t \in R_+} B(t). \qquad (5.30)$$

Let $x(t)$ be a solution of (5.1) on R_+. By integration of (5.1) we obtain

$$x(t) - x(0) = H(t) - \int_0^t \varphi(x(t - s))B(s)\, ds. \qquad (5.31)$$

For $t \geq T$ we can write

$$x(t) - x(0) + B(\infty)\int_0^t \varphi(x(s))\, ds$$

$$= H(t) + \left(\int_0^T + \int_T^t\right)\varphi(x(t - s))[B(\infty) - B(s)]\, ds. \qquad (5.32)$$

Taking into account condition (3), (5.28), and (5.30), we obtain for $t \geq T$

$$x(t) - x(0) + B(\infty)\int_0^t \varphi(x(s))\, ds$$

$$\geq \int_0^t \varphi(x(t - s))[B(\infty) - B(s)]\, ds - (\lambda M + H_0). \qquad (5.33)$$

The integral in the right side of (5.33) yields for $t \geq T$

$$\int_0^T \varphi(x(t - s))[B(\infty) - B(s)]\, ds = \int_0^T [\varphi(x(t - s)) + \lambda][B(\infty) - B(s)]\, ds$$

$$- \lambda \int_0^T [B(\infty) - B(s)]\, ds$$

$$\geq -B_0 \int_0^T [\varphi(x(t - s)) + \lambda]\, ds - \lambda M$$

$$= -B_0 \int_0^T \varphi(x(t - s))\, ds - \lambda(TB_0 + M).$$

Taking the extreme terms and considering (5.33), we get

$$x(t) - x(0) + B(\infty) \int_0^t \varphi(x(s))\, ds \geq -B_0 \int_{t-T}^t \varphi(x(s))\, ds - c, \qquad (5.34)$$

with $c = \lambda(TB_0 + 2M) + H_0$.

We claim now that there exists a positive constant c_1 such that

$$\int_0^t \varphi(x(s))\, ds \geq -c_1, \qquad t \in R_+. \qquad (5.35)$$

Indeed, if we assume that (5.35) does not hold, then we can find a sequence $\{t_n\}$, with $T < t_n < t_{n+1}$, $n = 1, 2, \ldots$, such that $t_n \to \infty$ and

$$\int_0^{t_n} \varphi(x(s))\, ds \to -\infty$$

as $n \to \infty$. Simple arguments show that there is no loss of generality if we assume that

$$\int_0^t \varphi(x(s))\, ds > \int_0^{t_n} \varphi(x(s))\, ds, \qquad 0 \leq t < t_n. \qquad (5.36)$$

From (5.36) we see that $\varphi(x(t_n)) \leq 0$. Taking into account condition (2), we obtain $x(t_n) \leq X$, $n = 1, 2, \ldots$. The inequality (5.34) leads to

$$X - x(0) + B(\infty) \int_0^{t_n} \varphi(x(s))\, ds$$

$$\geq -B_0 \left[\int_0^{t_n} \varphi(x(s))\, ds - \int_0^{t_n-T} \varphi(x(s))\, ds \right] - c \geq -c, \qquad (5.37)$$

if we also consider (5.36). But (5.37) and $B(\infty) > 0$ contradict

$$\int_0^{t_n} \varphi(x(s))\, ds \to -\infty$$

as $n \to \infty$. Therefore, (5.35) is established.

We shall now use (5.35) in order to prove

$$\limsup_{t \to \infty} x(t) < \infty. \qquad (5.38)$$

Let us remark (see the proof of (5.5)) that for $t \geq T$ we can write

$$x(t) - x(0) = H(t) - \left(\int_0^T + \int_T^t \right) \left\{ \int_0^{t-s} \varphi(x(u))\, du \right\} dB(s)$$

$$= H(t) - B(T) \int_0^{t-T} \varphi(x(s))\, ds - \int_0^T \varphi(x(t-s))B(s)\, ds$$

$$- \int_T^t \left\{ \int_0^{t-s} \varphi(x(u))\, du \right\} dB(s)$$

because an integration by parts yields

$$\int_0^T \left\{ \int_0^{t-s} \varphi(x(u))\, du \right\} dB(s) = B(s) \int_0^{t-s} \varphi(x(u))\, du \Big|_0^T - \int_0^T \varphi(x(t-s))B(s)\, ds.$$

We get further, on behalf of (5.35),

$$x(t) - x(0) \le H_0 + c_1 B(T) + \lambda \int_0^T B(s)\, ds + c_1 \int_T^t dB(s)$$

$$= H_0 + c_1 B(\infty) + \lambda \int_0^T B(s)\, ds, \tag{5.39}$$

for any $t \ge T$. (5.39) implies (5.38).

If we succeed in proving that

$$\liminf_{t \to \infty} x(t) > -\infty, \tag{5.40}$$

then Theorem 5.4 is proved in the case when $B(t)$ is nondecreasing on $[T, \infty)$. We follow the same lines as in the proof of Theorem 5.2, introducing the functions $\psi(y)$ and $y(t)$ by the formulas (5.17). The equation we find for $y(t)$ satisfies the assumptions of our theorem, and the arguments above lead to $\limsup y(t) < \infty$ as $t \to \infty$. This means that (5.40) holds true.

We shall now discuss the case when $B(t)$ is nonincreasing on $[T, \infty)$. Let $x(t)$ be a solution of (5.1) on R_+. It satisfies (5.31) and (5.32). If we integrate by parts in the right side of (5.32), we obtain

$$x(t) - x(0) + B(\infty) \int_0^t \varphi(x(s))\, ds = H(t) + \int_0^T \varphi(x(t-s))[B(\infty) - B(s)]\, ds$$

$$+ [B(\infty) - B(T)] \int_0^{t-T} \varphi(x(s))\, ds$$

$$- \int_T^t \left\{ \int_0^{t-s} \varphi(x(u))\, du \right\} dB(s), \tag{5.41}$$

for $t \ge T$. If we repeat the calculation that led to (5.34), we get from (5.41)

$$x(t) - x(0) + B(\infty) \int_0^t \varphi(x(s))\, ds \ge -B_0 \int_{t-T}^t \varphi(x(s))\, ds - c_2$$

$$+ [B(\infty) - B(T)] \int_0^{t-T} \varphi(x(s))\, ds - \int_T^t \left\{ \int_0^{t-s} \varphi(x(u))\, du \right\} dB(s), \tag{5.42}$$

with $c_2 = \lambda(TB_0 + M) + H_0$ and $t \ge T$. From (5.42) we can derive—as seen above—an inequality of the form (5.35). Indeed, if we suppose that

$$\int_0^t \varphi(x(s))\, ds$$

is not bounded from below, we can find a sequence $\{t_n\}$, $T < t_n < t_{n+1}$, $t_n \to \infty$ as $n \to \infty$, such that the integral tends to $-\infty$ on this sequence and (5.36) holds. One again obtains $x(t_n) \le X$. From

$$\int_0^{t_n - T} \varphi(x(s))\,ds = \int_0^{t_n} \varphi(x(s))\,ds - \int_{t_n - T}^{t_n} \varphi(x(s))\,ds$$

$$\le \int_0^{t_n} \varphi(x(s))\,ds + \lambda T$$

and $B(T) > B(\infty)$, we get

$$[B(\infty) - B(T)] \int_0^{t_n - T} \varphi(x(s))\,ds \ge [B(\infty) - B(T)] \int_0^{t_n} \varphi(x(s))\,ds$$

$$+ \lambda T[B(\infty) - B(T)]. \qquad (5.43)$$

We have further

$$-\int_T^{t_n} \left\{ \int_0^{t_n - s} \varphi x(u))\,du \right\} dB(s) \ge \left(\int_0^{t_n} \varphi(x(s))\,ds \right)\left(-\int_T^{t_n} dB(s) \right)$$

$$\ge [B(T) - B(\infty)] \int_0^{t_n} \varphi(x(s))\,ds. \quad (5.44)$$

Considering now (5.42)–(5.44) and $x(t_n) \le X$, we obtain

$$X - x(0) + B(\infty) \int_0^{t_n} \varphi(x(s))\,ds \ge -B_0 \int_{t_n - T}^{t_n} \varphi(x(s))\,ds - c_3, \quad (5.45)$$

where $c_3 = c_2 + \lambda T[B(T) - B(\infty)]$. The inequality (5.45) plays the same role as (5.37) and leads to a contradiction with our assumption that $\int_0^t \varphi(x(s))\,ds$ is not bounded from below. Therefore, an inequality of the form (5.35) holds true.

Starting now from

$$x(t) - x(0) = H(t) - \left(\int_0^T + \int_T^t \right) \varphi(x(t - s))B(s)\,ds$$

and using our hypothesis and (5.35), we find

$$x(t) - x(0) \le H_0 + \lambda \int_0^T B(s)\,ds - \int_T^t \varphi(x(t - s))[B(\infty) - B(s)]\,ds$$

$$- B(\infty) \int_T^t \varphi(x(t - s))\,ds \le H_0 + \lambda M + c_1 B(\infty) + \int_0^T B(s)\,ds$$

for $t \ge T$. From the above inequality we easily derive (5.38).

An inequality of the form (5.40) can be established if we again introduce the functions $\psi(y)$ and $y(t)$ by the formulas (5.17). Theorem 5.4 is thereby proven.

Remark The boundedness result established in Theorem 5.4 is actually stronger than stated above. In fact, we proved also that

$$\sup_{t \in R_+} \left| \int_0^t \varphi(x(s)) \, ds \right| < \infty. \tag{5.46}$$

Indeed, besides (5.35) we have to prove that $\int_0^t \varphi(x(s)) \, ds$ is bounded from above on R_+. But this results from the inequality similar to (5.35) for $\psi(y)$:

$$\int_0^t \psi(y(t)) \, dt \geq -c_1', \qquad t \in R_+,$$

with $c_1' > 0$.

In concluding this section we mention that all the results concerning the Volterra–Stieltjes equation (5.1) are due to Levin [4]. In his paper there are several other results pertaining to the boundedness or oscillation of the solutions, as well as interesting examples clarifying the role of various hypothesis.

5.6 Tauberian Results

An interesting tauberian result concerning the Volterra linear equation

$$x(t) = h(t) - \int_0^t k(t - s)x(s) \, ds, \qquad t \in R_+, \tag{6.1}$$

has been established long time ago by Paley and Wiener [1]. We shall state (and prove) this result in a slightly modified form due to Miller [2].

Theorem 6.1 (Paley–Wiener) Suppose that

$$h(t) \in C_{\ell}(R_+, R), \, k(t) \in L(R_+, R),$$

and

$$1 + \tilde{k}(s) \neq 0, \qquad s \in R. \tag{6.2}$$

If $x(t) \in L^\infty(R_+, R)$ is a solution of Eq. (6.1), then $x(t) \in C_{\ell}(R_+, R)$ and

$$x(\infty) = h(\infty) \Big/ \left(1 + \int_0^\infty k(t) \, dt\right). \tag{6.3}$$

Proof Since our conditions imply $x(t) \in C(R_+, R)$, we can use Lemma 4.1. The limit equation corresponding to (6.1) is

$$X(t) = h(\infty) - \int_{-\infty}^t k(t - s)X(s) \, ds, \qquad t \in R. \tag{6.4}$$

We shall now prove that (6.4) has a unique solution (in L^∞) that reduces to a constant. Indeed, Eq. (6.4) is of the form (2.14.1) with $k(t) = 0$ on $(-\infty, 0)$.

Condition (6.2) guarantees the existence and uniqueness of the solution in $L^\infty(R_+, R)$. It is very easy to check that this solution is given by $X(t) \equiv x(\infty)$, with $x(\infty)$ defined by (6.3).

From the fact that the positive limit set of $x(t)$ reduces to the unique point $x(\infty)$, there results $x(t) \to x(\infty)$ as $t \to \infty$. Therefore, $x(t) \in C_\ell(R_+, R)$ and its limit at infinity is given by (6.3).

A nonlinear version of Theorem 6.1 has been proved by Levin [2]. However, it should be noticed that it does not contain Theorem 6.1 as a special case, because conditions of a different nature are assumed on the kernel. Again following Miller [2], this result can be formulated in the form of the following theorem.

Theorem 6.2 Consider the equation

$$x(t) = h(t) - \int_0^t k(t - s)\varphi(x(s)) \, ds \tag{6.5}$$

on the half-axis R_+ and assume that the following conditions are verified:

1. $h(t) \in C_\ell(R_+, R)$;
2. $k(t) \in C(R_+, R) \cap L(R_+, R)$, is continuously differentiable on $(0, \infty)$, $k(t) \geq 0$, $\dot{k}(t) \leq 0$, and $\dot{k}(t) \not\equiv 0$ on any interval of R_+, except possibly $\dot{k}(t) \equiv 0$ for all sufficiently large t;
3. $\varphi(x)$ is continuous from R into itself, is strictly increasing, and such that $\varphi(0) = 0$.

Then any bounded solution $x(t)$ of (6.5) belongs to $C_\ell(R_+, R)$ and $x(\infty)$ is uniquely determined by the equation

$$x(\infty) + \varphi(x(\infty)) \int_0^\infty k(s) \, ds = h(\infty). \tag{6.6}$$

The proof of Theorem 6.2 depends upon the study of the limit equation corresponding to (6.5), namely,

$$X(t) = h(\infty) - \int_{-\infty}^t k(t - s)\varphi(X(s)) \, ds, \qquad t \in R. \tag{6.7}$$

If we consider the function

$$\phi(c) = c + \varphi(c) \int_0^\infty k(s) \, ds, \qquad c \in R,$$

we can easily see that there exists a unique value of c such that $\phi(c) = h(\infty)$, no matter we take $h(\infty) \in R$. In other words, the limit equation (6.7) has a constant solution. Using more technical arguments, one can prove that

$X(t) \equiv x(\infty)$, with $x(\infty)$ satisfying (6.6), is the only bounded solution on R for (6.7). The last property leads easily to the conclusion of Theorem 6.2.

Remark The fact that we omit the details of the proof does not mean that they are trivial. On the contrary, the proof of uniqueness requires a good deal of sharp arguments. The recent paper by Levin and Shea [1] contains the complete proof of Theorem 6.2 as well as many other tauberian results.

A significant achievement in applying tauberian methods to the investigation of the behavior of solutions for various classes of integral or integro-differential equations is due to Levin and Shea [1]. In concluding this section, we shall state a result of these authors generalizing Theorem 6.1.

Theorem 6.3 Consider the integral equation

$$x(t) + \int_{-\infty}^{\infty} x(t - s)\, dA(s) = h(t), \qquad t \in R, \tag{6.8}$$

under the following assumptions:

1. $h(t) \in L^{\infty}(R, R)$ and $\lim_{t \to \infty} h(t) = h(\infty)$ exists;
2. $A(t)$ is of bounded variation on R, left-continuous, and such that $A(-\infty) = 0$;
3. $x(t) \in L^{\infty}(R, R)$, is Borel measurable, satisfies (6.8), and

$$\lim |x(t + \tau) - x(t)| = 0 \qquad \text{as} \quad t \to \infty, \quad \tau \to 0.$$

If the equation

$$1 + \tilde{A}(\lambda) = 1 + \int_{-\infty}^{\infty} e^{i\lambda t}\, dA(t) = 0 \tag{6.9}$$

has no real solutions, then

$$\lim_{t \to \infty} x(t) = h(\infty)/(1 + A(\infty)). \tag{6.10}$$

If Eq. (6.9) has a finite set of real solutions, say $\{\lambda_1, \lambda_2, \ldots, \lambda_n\}$, then

$$x(t) = h(\infty)/(1 + A(\infty)) + \sum_{k=1}^{n} c_k(t) e^{-i\lambda_k t} + \eta(t), \tag{6.11}$$

where $\eta(t) \to 0$ as $t \to \infty$, and the $c_k(t)$ are bounded on R, of class C^{∞}, and such that $\lim c_k(t) = 0$ as $t \to \infty$, $k = 1, 2, \ldots, n$.

Remark Formula (6.3) is nothing but (6.10) when $A(t) \equiv 0$ on the negative half-axis. It can be easily seen that the condition occurring in (3) is automatically verified by any L^{∞} solution of the Volterra equation with integrable kernel.

For further details concerning tauberian results related to integral equation, we refer the reader to the paper by Levin and Shea [1]. It surveys all the significant results in this field and outlines an interesting research program.

Exercises

1. Consider the equation

$$\dot{x}(t) = - \int_0^t k(t-s)\varphi(x(s))\, ds, \qquad t \in R_+,$$

and assume that $\varphi(x)$ satisfies hypothesis H_4. If $k(t)$ is such that condition (1) of Lemma 2.1 holds, prove that $\lim x^{(j)}(t) = 0$ as $t \to \infty$, $j = 0, 1, 2$.

Hint: Use the function

$$E(t) = \phi(x(t)) + \tfrac{1}{2}k(t)\left(\int_0^t \varphi(x(s))\, ds \right)^2 - \tfrac{1}{2}\int_0^t k(t-s)\left[\int_s^t \varphi(x(u))\, du \right]^2 ds$$

or apply Lemma 2.1 directly.

2. Consider the kernel

$$k_1(t, s) = \gamma(t)\rho(s)k(t-s),$$

where $k(t)$ satisfies condition (1) of Lemma 2.1 and $\gamma(t)$ and $\rho(s)$ are such that the following conditions hold:

$$k(t) \in C(R_+, R), \qquad (-1)^j \gamma^{(j)}(t) \geq 0, \qquad 0 < t < \infty,$$

$$j = 0, 1, 2, \qquad \gamma(\infty) > 0, \qquad \rho(t), \dot{\rho}(t) \in C(R_+, R),$$

$$\rho^{(j)}(t) \geq 0, \qquad t \in R_+, \qquad j = 0, 1, \qquad 0 < \rho(\infty) < \infty.$$

Prove that $k_1(t, s)$ satisfies the hypotheses required in Theorems 1.1 and 1.2.

3. Show that the integro-differential equation

$$\dot{x}(t) = - \int_0^t x(s) \exp\{-(t+2)^2(t-s)\}\, ds, \qquad t \in R_+,$$

is not stable.

Hint: Let $x(t)$ be the solution with $x(0) = 1$. If we assume $x(t) \to 0$ as $t \to \infty$, then there exists T, $0 < T < \infty$, such that $x(T) = \tfrac{1}{4}$ and $x(t)$ decreases on $(0, T)$. Integrating both sides of the equation from 0 to T and using some elementary estimations, we get a contradiction. Therefore, $x(t)$ cannot go to zero as $t \to \infty$.

4. Consider the system

$$\dot{u}(t) = - \int_{-\infty}^{\infty} \alpha(x)T(x, t)\, dx + \sigma(t), \qquad T_t = T_{xx} + \eta(x)\varphi(u(t))$$

and apply the method developed in Section 5.2. Here, $\sigma(t) \in L(R_+, R)$ is the perturbation, and this setting allows the incorporation of some physical effects that were ignored in (2.1).

5. Consider the system

$$\dot{u}(t) = -\int_0^\pi \alpha(x)T(x, t) \, dx, \qquad T_t = T_{xx} + \eta(x)\varphi(u(t)),$$

$0 \le x \le \pi$, $0 < t < \infty$, and investigate the possibility of reduction to an integro-differential equation for $u(t)$, when more general boundary value conditions than those in Remark 1 to Theorem 2.1 are involved.

6. Consider the integro-differential system

$$\dot{x}(t) = m\varphi(x(t)) + \int_0^t k(t - s)\varphi(x(s)) \, ds, \qquad t \in R_+,$$

with the initial condition $x(0) = x^0 \in R^n$, m being a constant. Find the equivalent integral system and, using Theorem 4.1 and Remark 1 to this theorem, give conditions under which the trivial solutions is stable.

7. Consider the integral equation

$$x(t) = -\int_0^t \varphi(x(t - s))B(s) \, ds + F(t), \qquad t \in R_+,$$

with φ and B satisfying conditions (2)–(4) of Theorem 5.4, and $F(t) \in C(R_+, R)$. Then any solution of the above equation is bounded on R_+, if it exists there.

8. Investigate the integro-differential equation of infinite delay type

$$\dot{x}(t) = -\int_0^\infty \varphi(x(t - s)) \, dB(s) + f(t), \qquad t \in R_+.$$

The initial data is now a prescribed function $x_0(t) \in C(R_-, R)$. State and prove a boundedness result.

Bibliographical Notes

The paternity of the results included in this chapter was shown in each section. The exercises were selected from the papers of those authors to whom the results themselves belong. We shall now give further references on the topics discussed in this chapter.

The dynamics of nuclear reactors and the corresponding mathematical problems have been investigated by many authors. Among the first papers devoted to such topics and whose content is close to the spirit of this chapter, we shall mention those of Levin and Nohel [1, 2] and Smets [1, 2]. Using tauberian theorems for the Laplace transform, Levin and Nohel [1, 2] prove under suitable conditions that the solution of the system

$$\dot{u}(t) = -\int_{-\infty}^{\infty} \alpha(x)T(x, t)\, dx, \qquad T_t = T_{xx} + \eta(x)u,$$

with the initial conditions

$$u(0) = u_0, \qquad T(x, 0) = f(x), \qquad -\infty < x < \infty,$$

is such that $\lim t^{3/2}u(t) = $ constant as $t \to \infty$ and $T(x, t) = O(t^{-1/2})$ as $t \to \infty$. The results of Smets [1, 2] are generalized by those included in Section 5.5.

In his paper [3], Miller finds conditions under which the system

$$\dot{u}(t) = -\int_{0}^{\pi} \alpha(x)T(x, t)\, dx, \qquad T_t = T_{xx} + \eta(x)\varphi(u)$$

is unstable. The boundary value condition is of the form $T_x(0, t) = T_x(\pi, t)$, i.e., a periodicity condition.

Another paper of Miller [5] contains an application of the linearization theorem to the investigation of stability of a system occurring in reactor dynamics. Finally, the almost-periodic asymptotic behavior of solution is emphasized by Miller in his paper [6].

More intricate systems describing various phenomena in reactor dynamics have been recently investigated by London [1–3]. His conspicuous paper [1] is partly dominated by the method of Levin we presented in Section 5.5. The last part of the paper contains applications of the Leray–Schauder degree theory to some nonlinear Volterra integral equations. The following result is established by London in [3]. Consider the system

$$\dot{x}(t) = h(t) + \int_{0}^{t} k(t - s)\varphi(x(s))\, ds + y(t)/\varphi(x(t)),$$

$$\dot{y}(t) = \alpha\varphi(x(t)) - \beta y(t),$$

and let $x(t)$, $y(t)$ be a solution on R_+ such that $x(0)$ is finite and $y(0) > 0$. Assume that the following conditions hold: (1) $h(t) \in C(R_+, R)$ and

$$h(t) - H \in L(R_+, R)$$

for some $H \geq 0$; (2) $\varphi(x)$ is a continuously differentiable map of R into itself with $\varphi(x) > 0$, $\varphi'(x) \geq 0$, $\lim \varphi(x) = 0$ as $x \to -\infty$, and $\lim \varphi(x) = \infty$ as $x \to \infty$; moreover, there exists a positive number γ such that $\gamma\varphi'(x) \leq \varphi(x)$ on R; (3) $k(t) \in L(R_+, R)$ and is continuously differentiable of the second order on R_+, $k(t) \leq 0$, $k(0) < 0$, $\dot{k}(t) \leq -\rho k(t)$, with $\rho = \min\{\beta^{-1}, \gamma/H\}$ if $H > 0$ and $\rho = \beta^{-1}$ if $H = 0$. Then $\lim \varphi(x(t))$ and $\lim y(t)$ exist as $t \to \infty$ and satisfy

$$\beta \lim_{t \to \infty} y(t) = \alpha \lim_{t \to \infty} \varphi(x(t)) = (\alpha/\beta + H) \int_{0}^{\infty} |k(s)|\, ds.$$

The integral equation

$$x(t) = h(t) + \int_0^t k(t - s)\varphi(s, x(s))\, ds, \qquad t \in R_+,$$

has been considered, under various assumptions, by Friedman [1, 2]. In particular, the existence of asymptotically periodic solutions is established. Some results generalizing those of Friedman have been obtained by Miller [4]. A basic property occurring in Miller's paper [4] is that of positivity of the resolvent kernel associated with $k(t)$.

Some boundedness results for Volterra–Stieltjes integral equations have been proved by Petrovanu [2].

An interesting tauberian result for the Volterra's population equation

$$(\dot{N}(t)/N(t)) = a - bN(t) - \int_c^t k(t - s)N(s)\, ds,$$

where $c = 0$ or $c = -\infty$, has been obtained by Miller [1]. Under natural conditions, suggested by the interpretation of the equation as modeling the growth of a biological population, it is shown that $N(\infty) = a[b + \int_0^\infty k(s)\, ds]^{-1}$. Properties related to the positive limit set of a bounded solution are involved.

Hannsgen [1–3] applies various tauberian theorems to the investigation of the asymptotic behavior of solutions for various classes of integral or integro-differential equations. In [3], the equation $\dot{u}(t) = -\int_0^t k(t - s)u(s)\, ds$ with completely monotonic kernel (i.e., such that $(-1)^j k^{(j)}(t) \geq 0$) is investigated in this way. The basic result can be extended also to some nonlinear equations.

In concluding these notes we shall make a few comments with respect to the general problems of local existence and continuation of solutions. We had several opportunities in this chapter to invoke such results for integral equations. The recent book by Miller [7], the paper by Neustadt [1], as well as the papers by Nohel [1] and Azbelev and Calyuk [1] furnish the needed results or the necessary tools to prove them without difficulty.

References

M. A. Aizerman and F. R. Gantmacher
[1] "Absolute Stability of Regulator Systems." Holden-Day, San Francisco, 1963.

S. Albertoni, A. Cellina, and G. P. Szegö
[1] Comportamento asintotico di una equazione integrale non lineare, *Ann. Mat. Pura Appl.* **72** (1966), 133–140.

S. Albertoni and G. Szegö
[1] Sur une équation intégrale de la théorie du réglage, *C. R. Acad. Sci. Paris* **261** (1965), Groupe 1, 29–32.

C. Avramescu
[1] Sur l'existence des solutions des équations intégrales dans certains espaces fonctionnels, *Ann. Univ. Sci. Budapest. Eötvös Sect. Math.* **13** (1970), 19–34.

N. V. Azbelev and Z. B. Calyuk
[1] On integral inequalities, I. (Russian), *Mat. Sb.* **56** (1962), 325–342.

R. A. Baker and D. J. Vakharia
[1] Input–output stability of linear time-invariant systems, *IEEE Trans. Automatic Control* **AC-15** (1970), 316–319.

G. Bantaş
  [1] Contributions to the study of integral equations (Romanian), Ph.D. thesis, University of Iaşi, Romania, 1970.

I. Barbălat
  [1] Systèmes d'équations différentielles d'oscillations non-linéaires, *Rev. Roumaine Math. Pures Appl.* **4** (1959), 267–270.

I. Barbălat and A. Halanay
  [1] Hyperstabilité des blocs intégraux et à différences, *Rev. Roumaine Sci. Tech. Sér. Électrotechnique et Énergétique* **14** (1969), 631–637.

V. Barbu
  [1] Sur une équation intégrale non-linéaire, *An. Şti. Univ. "Al. I. Cuza" Iaşi Secţ. Ia Mat.* **X** (1964), 61–65.

R. Bellman
  [1] On the application of a Banach–Steinhaus theorem to the study of the boundedness of solutions of non-linear differential and difference equations, *Ann. of Math.* **49** (1948), 515–522.
  [2] "Introduction to the Mathematical Theory of Control Processes," vol. 1. Academic Press, New York, 1967.

R. Bellman and K. Cooke
  [1] "Differential-Difference Equations." Academic Press, New York, 1963.

V. E. Beneš
  [1] A nonlinear integral equation from the theory of servomechanisms, *Bell System Tech. J.* **40** (1961), 1309–1321.
  [2] A fixed point method for studying the stability of a class of integro-differential equations, *J. Math. and Phys.* **40** (1961), 55–67.
  [3] A nonlinear integral equation in the Marcinkiewicz space $\mathcal{M}_2$, *J. Math. and Phys.* **XLIV** (1965), 24–35.

I. Bihari
  [1] Notes on a nonlinear integral equation, *Studia Sci. Math. Hungar.* **2** (1967), 1–6.

R. E. Blodgett and R. E. King
  [1] Absolute stability of a class of nonlinear systems containing distributed elements *J. Franklin Inst.* **284** (1967), 153–160.
  [2] Quasi-asymptotic stability of a class of non-linear systems, *Internat. J. Control* **8** (1968), 245–252.

N. Bourbaki
  [1] "Élements de Mathématique: Livre III, Topologie Générale." Hermann, Paris, 1958, Ch. IX, X.

T. F. Bridgland
  [1] *IRE Trans. Circuit Theory* **CT-10** (1963), 539–542.

R. W. Brockett
  [1] "Finite Dimensional Linear Systems." Wiley, New York, 1970.

R. W. Brockett and J. L. Willems
  [1] Frequency domain stability criteria, I, II. *IEEE Trans. Automatic Control* **10** (1965), 255–261, 407–413.

F. E. Browder
[1] Non-linear equations of evolution, *Ann. of Math.* **80** (1964) 485–523.

M. S. Budjanu
[1] On the solution of some classes of Wiener–Hopf equations with operator coefficients, *Izv. Akad. Nauk Moldav. SSR.* (1966), 18–31.

Z. B. Calyuk
[1] On stability of Volterra equation (Russian), *Differencial'nye Uravnenija* **IV** (1968), 1967–1979.

G. N. Čebotarev
[1] On rings whose elements are functions integrable with respect to a weight function (Russian), *Izv. Vysš Ucebn. Zaved. Matematika* (1963), **36**, 133–145.

G. N. Čebotarev and V. B. Govorova
[1] On the theory of convolution equations with negative index (Russian), *Trudy Kazan. Aviacion. Inst.* **89** (1965), 109–115.

I. S. Čebotaru
[1] On approximative methods in solving integral equations of Wiener–Hopf type (Russian), *Issled. Alg. Mat. An.* (1965), 79–96, Kishinev.

Y. S. Cho and K. S. Narendra
[1] An off-axis circle criterion for the stability of feedback systems with monotonic nonlinearity, *IEEE Trans. Automatic Control* **13** (1968), 413–416.

J. Chover and P. Ney
[1] A nonlinear integral equation and its application to critical branching processes. *J. Math. Mech.* **14** (1965), 723–735.
[2] The non-linear renewal equation, *J. Analyse Math.* **XXI** (1968), 381–413.

R. Conti
[1] On the boundedness of solutions of ordinary differential equations, *Funkcial. Ekvac.* **9** (1966), 23–26.

W. A. Coppel
[1] "Stability and Asymptotic Behavior of Differential Equations." Heath, Boston, Massachusetts, 1965.

A. Corduneanu
[1] Le spectre de l'opérateur de Volterra dans certains espaces de fonctions, *Bul. Inst. Politehn. Iași* **XII** (1967), 77–81.
[2] Sur le spectre de l'opérateur de Volterra, *Bul. Inst. Politehn. Iași* **XIV** (1968), 79–82.
[3] On stability of solutions of singular integral equations of Volterra type (Russian), *Bul. Inst. Politehn. Iași* **XII** (1968), 31–34.
[4] Équations Volterra et certaines applications, *Bul. Inst. Politehn. Iași* **XVI** (1970), 37–45.

C. Corduneanu
[1] Sur une équation intégrale de la théorie du réglage automatique, *C. R. Acad. Sci. Paris* **256** (1963), 3564–3567.
[2] Sur une équation intégrale non-linéaire, *An. Şti. Univ. "Al. I. Cuza Iași" Sect. Ia Mat.* **IX** (1963), 369–375.
[3] Problèmes globaux dans la théorie des équations intégrales de Volterra, *Ann. Mat. Pura Appl.* **LXVII** (1965), 349–363.

[4] Quelques problèmes qualitatifs de la théorie des équations intégro-différentielles. *Colloq. Math.* **18** (1967), 77–87.

[5] Some perturbation problems in the theory of integral equations, *Math. Systems Theory* **I** (1967), 143–155.

[6] Nonlinear perturbed integral equations, *Rev. Roumaine Math. Pures Appl.* **XIII** (1968), 1279–1284.

[7] On a class of functional-integral equations. *Bull. Math. Soc. Sci. Math. R.S. Roumanie.* **12** (1968), 43–53.

[8] "Almost Periodic Functions." Wiley, New York, 1968.

[9] Stability of some linear time-varying systems, *Math. Systems Theory* **III** (1969), 151–155.

[10] Admissibility with respect to an integral operator and applications, *SIAM Studies in Appl. Math.* **5** (1969), 55–63.

[11] Periodic and almost periodic solutions of some convolution equations. *Proc. (Trudy) 5th Int. Conf. Nonlinear Oscillations*, vol. I, 311–320. Kiev, 1970.

[12] "Principles of Differential and Integral Equations." Allyn and Bacon, Boston Massachusetts, 1971.

J. Cronin
[1] Fixed points and topological degree in nonlinear analysis, *Math. Surveys* Amer. Math. Soc., Providence, Rhode Island, 1964.

C. M. Dafermos
[1] An abstract Volterra Equation with applications to linear viscoelasticity, *J. Differential Equations* **7** (1970), 554–569.

C. A. Desoer
[1] A generalization of the Popov criterion, *IEEE Trans. Automatic Control* **AC-10** (1965), 182–185.

[2] A general formulation of the Nyquist criterion, *IEEE Trans. Circuit Theory* **CT-12** (1965), 230–234.

C. A. Desoer and A. J. Thomasian
[1] A note on zero-state stability of linear systems, *Proc. 1st Allerton Conf. on Circuit and System Theory*, Urbana, Illinois, 1963, 50–52.

C. A. Desoer and M. Y. Wu
[1] Stability of linear time-invariant systems, *IEEE Trans. Circuit Theory* **CT-15** (1968), 245–250.

[2] Stability of multiple-loop feedback linear time-invariant systems, *J. Math. Anal. Appl.* **23** (1968), 121–129.

[3] Stability of a nonlinear time-invariant feedback systems under almost constant inputs, *Automatica* **5** (1969), 231–233.

A. G. Dewey
[1] On the stability of feedback systems with one differentiable nonlinear element, *IEEE Trans. Automatic Control* **AC-11** (1966), 485–491.

V. Doležal
[1] An extension of Popov's method for vector-valued nonlinearities, *Czechoslovak Math. J.* **15** (1965), 436–453.

[2] On general feedback systems containing delayors, *Aplikace Matematiky* **13** (1968), 489–507.

[3] On general nonlinear and quasilinear unanticipative feedback systems, *Apl. Mat.* **14** (1969), 220–240.

[4] A remark on energetic stability of feedback systems, *Apl. Mat.* **14** (1969), 345–354.

N. Dunford and J. T. Schwartz
[1] "Linear Operators, Part I." Wiley (Interscience), New York, 1958.

I. A. Feldman and I. C. Gochberg
[1] Integral-difference equations of Wiener–Hopf type (Russian), *Acta. Sci. Math. (Szeged)* **XXX** (1969), 199–224.

M. I. Freedman, P. L. Falb, and G. Zames
[1] A Hilbert space stability theory over locally compact Abelian groups, *SIAM J. Control* **7** (1969), 479–495.

A. Friedman
[1] On integral equations of Volterra type, *J. Analyse Math.* **11** (1963), 381–413.
[2] Periodic behavior of solutions of Volterra integral equations, *J. Analyse Math.* **15** (1965), 187–203.

A. Friedman and M. Shinbrot
[1] Volterra integral equations in Banach space, *Trans. Amer. Math. Soc.* **126** (1967), 131–179.

A. Kh. Geleg
[1] The absolute stability of non-linear control systems with distributed parameters *Automat. Remote Control* **26** (1965), 401–409.
[2] The absolute stability of nonlinear control systems with distributed parameters in critical cases, *Automat. Remote Control* **27** (1966).

I. Gel'fand, D. Raikov, and G. E. Shilov
[1] "Commutative Normed Rings." Chelsea, New York, 1964.

I. C. Gochberg
[1] The factorization problem in normed rings, functions of isometric and symmetric operators and singular integral equations (Russian). *Uspehi Mat. Nauk.* **19** (1964), 71–124.

I. C. Gochberg and I. A. Feldman
[1] "Convolution Equations and Projection Methods of Solving Them" (Russian). Nauka, Moscow, 1971.

I. C. Gochberg and M. G. Krein
[1] Systems of integral equations on a half line with kernels depending on the difference of arguments, *Amer. Math. Soc. Transl.* **14**, 217–287.

E. I. Goldenhershel
[1] The spectrum of a Volterra operator on the half-axis and the exponential growth of solutions of Volterra integral systems (Russian), *Mat. Sb.* **64** (1964), 115–139.

H. E. Gollwitzer
[1] Admissibility and integral operators, *Math. Systems Theory* (to appear).

S. I. Grossman
[1] Existence and stability of a class of nonlinear Volterra integral equations, *Trans. Amer. Math. Soc.* **150** (1970), 541–556.

S. I. Grossman and R. K. Miller
  [1] Perturbation theory of a Volterra Integro-differential Systems, *J Differential Equations* **8** (1970), 457–474.

A. Halanay
  [1] Absolute stability of some nonlinear regulator systems with time lag, *Automat. Remote Control* **25** (1964), 290–301.
  [2] On the asymptotic behavior of the solutions of an integro-differential equations, *J. Math. Anal. Appl.* **10** (1965), 319–324.
  [3] Asymptotic behavior of solutions of some nonlinear integral equations (Russian), *Rev. Roumaine Math. Pures Appl.* **X** (1965), 765–777.
  [4] Differential Equations: Stability, Oscillations, Time Lags. Academic Press, New York, 1966.
  [5] Almost periodic solutions for a class of nonlinear systems with time lag, *Rev. Roumaine Math. Pures Appl.* **XIV** (1969), 1269–1276.

J. K. Hale
  [1] "Functional Differential Equations." Springer Verlag, New York, 1971.

C. D. Han and A. U. Meyer
  [1] Stability analysis of a class of nonlinear distributed parameter system-tubular chemical reactors, *Internat. J. Control* **11** (1970), 509–528.

K. B. Hannsgen
  [1] Indirect Abelian theorems and a linear Volterra equation, *Trans. Amer. Math. Soc.* **142** (1969), 539–555.
  [2] On a nonlinear Volterra equation, *Michigan Math. J.* **16** (1969), 365–376.
  [3] A Volterra equation with completely monotonic convolution kernel, *J. Math. Anal. Appl.* **31** (1970), 459–471.

Ph. Hartman
  [1] "Ordinary Differential Equations." Wiley, New York, 1964.

M. Hukuhara
  [1] Sur l'existence des points invariants d'une transformation dans l'espace fonctionnel, *Japan J. Math.* **20** (1950), 1–4.

R. E. Kalman
  [1] *IRE Trans.* Circuit Theory, **CT-10** (1963), 420–422.

M. A. Krasnoselski
  [1] "Topological Methods in the Theory of Nonlinear Integral Equations." Pergamon Press, New York, 1964.

M. G. Krein
  [1] Convolution integral equations on the half-axis (Russian), *Uspehi Mat. Nauk.* **13** (1958), 3–120.

S. G. Krein (Editor)
  [1] "Functional Analysis" (Russian). Nauka, Moscow, 1964.

Y. H. Ku and H. T. Chieh
  [1] Extension of Popov's theorems for stability of nonlinear control systems, *J. Franklin Inst.*, **279** (1965), 401–416.

J. Kudrewicz
  [1] Application of the functional analysis methods for the investigation of stability of nonlinear electrical systems (Polish; English summary), *Rozprawy Elektrotech.* **IX** (1963), 3–50.

[2] Stability of feedback nonlinear systems, *Automat. Remote Control* **25** (1964), 1145–1155.
[3] Positive operators and conditions of stability of dynamical systems (Russian), *Bull. Polish Acad. Sci.* (Technical Sciences series) **XII** (1964), 921–924.
[4] Periodic oscillations in nonlinear systems (Polish), *Arch. Automat. i Telemech.* **XI** (1966), 373–389.
[5] "Frequency Methods in the Theory of Nonlinear Dynamical Systems" (Polish). Wydawnictwa Naukowo-Techniczne, Warszawa, 1970.

V. Lakshmikantham and S. Leela
[1] "Differential and Integral Inequalities," vol. I. Academic Press, New York, 1969.

S. Lefschetz
[1] "Stability of Nonlinear Control Systems." Academic Press, New York, 1965.

G. Lellouche
[1] A frequency criterion for oscillatory solutions, *SIAM J. Control* **8** (1970), 202–206.

J. J. Levin
[1] The asymptotic behavior of the solution of a Volterra equation, *Proc. Amer. Math. Soc.* **14** (1963), 534–541.
[2] The qualitative behavior of a nonlinear Volterra equation, *Proc. Amer. Math. Soc.* **16** (1965), 711–718.
[3] A nonlinear Volterra equation not of convolution type, *J. Differential Equations* **4** (1968), 176–186.
[4] Boundedness and Oscillation of some Volterra and delay equations, *J. Differential Equations* **5** (1969), 369–398.

J. J. Levin and J. A. Nohel
[1] On a system of integrodifferential equations occurring in reactor dynamics, *J. Math. Mech.* **9** (1960), 347–368.
[2] On a system of integrodifferential equations occurring in reactor dynamics II, *Arch. Rational Mech. Anal.* **11** (1962), 210–243.
[3] Note on a nonlinear Volterra equation, *Proc. Amer. Math. Soc.* **14** (1963), 924–929.
[4] On a nonlinear delay equation, *J. Math. Anal. Appl.* **8** (1964), 31–44.
[5] Perturbations of a nonlinear Volterra equation, *Michigan Math. J.* **12** (1965), 431–444.
[6] A system of nonlinear integro-differential equations, *Michigan Math. J.* **13** (1966), 257–270.
[7] A nonlinear system of integrodifferential equations, *in* "Mathematical Theory of Control" (proc. of a conference held at the Univ. Southern California, January 30–February 1, 1967). Academic Press, New York, 1967, 398–405.
[8] The integro-differential equations of a class of nuclear reactors with delayed neutrons, University of Wisconsin, Madison, Wisconsin, 1968.

J. J. Levin and D. F. Shea
[1] On the asymptotic behavior of the bounded solution of some integral equations. University of Wisconsin, Madison, Wisconsin, 1970.

S. O. Londen
[1] On some nonlinear Volterra equations, *Ann. Acad. Sci. Fenn. Ser. A VI*, no. 317, 1969.

[2]   On a nonlinear Volterra integrodifferential equation, *Comment. Physico-Math.* **38**, no. 2, 1969.
[3]   On the asymptotic behavior of the solution of a nonlinear integrodifferential equation, *SIAM J. Math. Analysis* **2** (1971), 356–367.

N. Luca
[1]   Sur quelques systèmes d'équations intégrales à noyau transitoire qui s'appliquent aux problèmes de réglage automatique, *An. Şti. Univ. "Al. I. Cuza" Iaşi Sect Ia Mat.* **X** (1964), 347–355.
[2]   On the behaviour of the solution of an integro-differential system of equations *An. Şti. Univ. "Al. I. Cuza" Iaşi Secţ. Ia Mat.* **XIII** (1967), 299–303.

J. L. Massera and J. J. Schäffer
[1]   Linear Differential Equations and Function Spaces. Academic Press, New York, 1966.

R. K. Miller
[1]   On Volterra's population equation, *SIAM J. Appl. Math.* **14** (1966), 446–452.
[2]   Asymptotic behavior of solutions of nonlinear Volterra equations, *Bull. Amer. Math. Soc.* **72** (1966), 153–156.
[3]   An unstable nonlinear integro-differential system, *Proc. U.S.–Japan Seminar on Differential and Functional Equations*, June 26–30, 1967, Benjamin, New York.
[4]   On Volterra integral equations with nonnegative integrable resolvents, *J. Math. Anal. Appl.* **22** (1968), 319–340.
[5]   On the linearization of Volterra integral equations, *J. Math. Anal. Appl.* **23** (1968), 198–208.
[6]   Almost periodic behavior of solutions of a nonlinear Volterra system, *Quart. Appl. Math.* **XXIX** (1971), 553–570.
[7]   "Nonlinear Volterra Integral Equations." Benjamin, New York, 1971.

R. K. Miller, J. A. Nohel, and J. S. W. Wong
[1]   A stability theorem for nonlinear mixed integral equations, *J. Math. Anal. Appl.* **25** (1969), 446–449.

R. K. Miller and G. R. Sell
[1]   Existence, uniqueness and continuity of solutions of integral equations, *Ann. Mat. Pura Appl.* **LXXX** (1968), 135–152.

J. Moser
[1]   On nonoscillating networks, *Quart. Appl. Math.* **XXV** (1967), 1–9.

N. I. Muskhelishvili
[1]   "Singular Integral Equations" (Russian). OGIZ Moscow–Leningrad, 1962.

B. N. Naumov and Ya. Z. Tsypkin
[1]   A frequency criterion for absolute stability of processes in nonlinear automatic control systems, *Automat. Remote Control* **25** (1964), 765–778.

L. W. Neustadt
[1]   On the solutions of certain integral-like operator equations. Existence, uniqueness and dependence theorems, *Arch. Rational Mech. Anal.* **38** (1970), 131–160.

B. Noble
[1]   "Methods Based on the Wiener–Hopf Technique for the Solution of Partial Differential Equations." Pergamon, London, 1958.

J. A. Nohel
  [1] Some problems in nonlinear Volterra integral equations, *Bull. Amer. Math. Soc.* **64** (1962), 323–329.
  [2] Problems in qualitative behavior of solutions of nonlinear Volterra equations, *in* "Nonlinear Integral Equations." Univ. Wisconsin Press, Madison, Wisconsin, 1964.
  [3] Qualitative behaviour of solutions of nonlinear Volterra equations, *in* "Stability Problems of Solutions of Differential Equations." Edizioni Oderisi, Gubbio, 1966.
  [4] Remarks on nonlinear Volterra equations, *Proc. U.S.–Japan Seminar on Differential and Functional Equations*, June 26–30, 1967, Benjamin, New York.

R. O'Neil
  [1] Integral transforms and tensor products on Orlicz spaces and $L(p, q)$ spaces, *J. Analyse Math.* **XXI** (1968), 5–276.

R. P. O'Shea
  [1] A combined frequency-time domain stability criterion for autonomous continuous systems, *IEEE Trans. Automatic Control* **AC-11** (1966). 177–184.

K. Padmavally
  [1] On a nonlinear integral equation, *J. Math. Mech.* **7** (1958), 533–555.

R. Paley and N. Wiener
  [1] Fourier transforms in the complex domain, American Math. Society, New York, 1934.

N. Pavel
  [1] Sur quelques problèmes de comportement global et de perturbation dans la théorie des équations intégrales non-linéaires de Volterra, *An. Şti. Univ. "Al. I. Cuza" Iaşi Secţ. Ia Mat.* **XVI** (1970), 315–325.

D. Petrovanu
  [1] Equations Hammerstein intégrales et discrètes, *An. Mat. Pura Appl.* **70** (1966), 227–254.
  [2] Solutions bornées pour des systèmes linéaires discrets et du type de Volterra–Stieltjes, *Bul. Inst. Politehn. Iaşi,* **XV (XIX)** (1969), 59–68.

V. M. Popov
  [1] Criterii de stabilitate pentru sistemele neliniare de reglare automată bazate pe utilizarea transformatei Laplace, *Stud. Cerc. Energetică* **IX** (1959), 119–135.
  [2] New graphical criteria for the stability of the steady state of nonlinear control systems, *Rev. d'Electrotechnique et d'Energétique* **VI** (1961), 25–34.
  [3] Absolute stability of nonlinear systems of automatic control, *Automat. Remote Control* **22** (1961), 857–875.
  [4] A critical case of absolute stability, *Automat. Remote Control* **23** (1962), 1–21.
  [5] "Hiperstabilitatea Sistemelor Automate." Editura Academiei R.S.R., Bucharest, 1966.

A. N. V. Rao
  [1] On some systems of automatic control theory, *An. Şti. Univ. "Al. I. Cuza" Iaşi Secţ. Ia Mat.* **XV** (1969), 47–57.
  [2] Stability of multivariable nonlinear systems containing distributed elements, *Bul. Inst. Politehn. Iaşi, Secţia I,* **XV (XIX),** (1969), 49–57.

I. M. Rapoport
   [1] On a class of singular integral equations (Russian), *Dokl. Akad. Nauk SSSR* **59**
        (1948), 1403–1406.

M. Reghiş
   [1] On non-uniform asymptotic stability, *J. Appl. Math. Mech.* **27** (1963), 344–362.

R. Reissig, G. Sansone, and R. Conti
   [1] "Nichtlineare Differentialgleichungen Höherer Ordnung." Ed. Cremonese, Roma,
        1969.

E. Reissner
   [1] On a class of singular integral equations, *J. Math. and Physics* **20** (1941), 219–223.

W. Rudin
   [1] "Real and Complex Analysis." McGraw-Hill, New York, 1966.

I. W. Sandberg
   [1] On the L_2-boundedness of solutions of nonlinear functional equations, *Bell System
        Tech. J.* **43** (1964), 1581–1600.
   [2] A frequency-domain condition for the stability of feedback systems containing a
        single time-varying nonlinear element, *Bell System Tech. J.* **43** (1964), 1601–1608.
   [3] On the boundedness of solutions of nonlinear integral equations, *Bell System Tech. J.*
        XLIV (1965), 439–453.
   [4] Some results on the theory of physical systems governed by nonlinear functional
        equation, *Bell System Tech. J.* **XLIV** (1965), 871–898.
   [5] Some stability results related to those of V. M. Popov, *Bell System Tech. J.* **XLIV**
        (1965), 2133–2148.

H. B. Smets
   [1] A general property of boundedness for the power of some stable and unstable
        nuclear reactors, *Nukleonik* **2** (1960), 44–45.
   [2] "Problems in Nuclear Power Reactor Stability." Presses Univ. Bruxelles, Brussels,
        Belgium, 1961.

A. Strauss
   [1] A discussion of the linearization of Volterra integral equations, *in* "Lectures Notes
        in Mathematics," vol. 144 (1970), Springer, Berlin, 209–217.
   [2] On a perturbed Volterra integral equation, *J. Math. Anal. Appl.* **30** (1970), 564–575.

G. P. Szegö
   [1] Sul comportamento asintotico di una equazione integrale non lineare, *An. Şti.
        Univ. "Al. I. Cuza" Iaşi, Secţ. Ia Mat.* **XV** (1969), 387–394.

J. H. Taylor and K. S. Narendra
   [1] The Corduneanu–Popov approach to the stability of nonlinear time-varying
        systems, *SIAM J. Appl. Math.* **18** (1970), 267–281.

C. P. Tsokos
   [1] On a stochastic integral equation of Volterra type, *Math. Systems Theory* **3** (1969),
        222–231.

F. A. Valentine
   [1] "Convex Sets." McGraw-Hill, New York, 1964.

V. R. Vinokurov
[1] On the stability of solutions of Volterra integral systems of the second kind. I, II (Russian), *Izv. Vysš. Učebn. Zaved. Matematika* (1959), 23–34, 50–58.

[2] A method in studying the asymptotic properties of the resolvent of an integral system of Volterra type (Russian), *Izv. Vysš. Učebn. Zaved. Matematika* (1964) 24–31.
[3] Some problems of the theory of stability for Volterra integral systems. I, II (Russian), *Izv. Vysš. Učebn. Zaved. Matematika* (1969), no. 6, 24–34, no. 7, 28–38.

V. Volterra
[1] Sur la théorie mathématique des phénomènes héréditaires, *J. Math. Pures Appl.* 7 (1928), 249–298.

N. Wiener and E. Hopf
[1] Uber eine Klasse singulärer Integralgleichungen, *Sitz. Akad. Wiss.* (Berlin) (1931), 696–708.

J. C. Willems
[1] Stability, instability, invertibility and causality, *SIAM J. Control* 7 (1969), 645–671.
[2] Some results on the L^p-stability of linear time-varying systems, *IEEE Trans. Automatic Control* AC-14 (1969), 660–665.
[3] " The Analysis of Feedback Systems," Research Monograph no. 62. M.I.T. Press, Cambridge, Massachusetts, 1971.

J. L. Willems
[1] A general stability criterion for non-linear time-varying feedback systems, *Internat. J. Control* 11 (1970), 625–631.

D. Willett
[1] Nonlinear vector integral equations as contraction mappings, *Arch. Rational Mech. Anal.* 15 (1964), 79–86.

M. Y. Wu and C. A. Desoer
[1] L^p-stability $(1 \leq p \leq \infty)$ of nonlinear time-varying feedback systems, *SIAM J. Control* 7 (1969), 356–364.

V. A. Yakubovitch
[1] Frequency conditions for the stability of solutions of nonlinear integral equations of automatic control (Russian), *Vestnik Leningrad. Univ.* 1967, 109–125.

K. Yosida
[1] " Functional Analysis." Springer, Berlin and New York, 1965.

D. C. Youla
[1] On the stability of linear systems, *IEEE Trans. Circuit Theory* CT-10 (1963), 276–279.

P. P. Zabreiko, A. I. Koshelev, M. A. Krasnoselski, S. G. Michlin, L. S. Rakovscik, and V. I. Stecenko
[1] " Integral Equations," (Russian). Nauka, Moscow, 1968.

L. A. Zadeh and C. A. Desoer
[1] " Linear Systems Theory." McGraw-Hill, New York, 1963.

G. Zames
  [1] Functional analysis applied to nonlinear feedback systems, *IEEE Trans. Circuit Theory* **CT-10** (1963), 392–404.
  [2] On the input-output stability of time-varying nonlinear feedback systems, I, II, *IEEE Trans. Automatic Control* **AC-11** (1966), 228–238, 465–476.
G. Zames and P. L. Falb
  [1] On the stability of systems with monotone and odd monotone nonlinearities, *IEEE Trans. Automatic Control* **AC–12** (1967), 221–223.
  [2] Stability conditions for systems with monotone and slope-restricted nonlinearities, *SIAM J. Control*, **6** (1968), 89–109.

Author Index

Subject Index

A

A (condition for a kernel), 28
A_ω, 3
$AP(R, R^n)$, 3–4
Absolute stability, 90, 95, 119
Admissibility, 24
  differential systems, 49–51
  integral operators, 24–28, 33, 34,
    37–39, 42, 46–48

B

B (condition for a kernel), 34
B_p (condition for a kernel), 37
Barbălat's lemma, 89, 93, 99, 104

C

$\mathscr{C}$ (the complex number field), 12
$C(R_+, R^n)$, 3
$C_c(R_+, R^n)$, 1–2
$C_g(R_+, R^n)$, 2

$C_\ell(R_+, R^n)$, 3
$C_0(R_+, R^n)$, 3
Contraction mapping principle, 11
Control systems
  automatic, 83, 84
  time lag, 116
Convolution equation
  linear, 66–70, 213, 215
  nonlinear, 83, 85, 86, 96, 101, 105, 112,
    117, 119, 124, 130, 131, 134, 136, 196
  solutions in L^2, 112–115
Convolution product, 14, 16

D

D-basis, 152
  existence of, 152, 159

E

Electric network, 109
Energetic stability, 125
  circle condition of, 129, 130